# THE
# UNIFICATION
# THEORY

# THE
# UNIFICATION
# THEORY

## VOLUME ONE

### KOK FAH CHONG

PARTRIDGE

Print information available on the last page.

**To order additional copies of this book, contact**
Toll Free 800 101 2657 (Singapore)
Toll Free 1 800 81 7340 (Malaysia)
orders.singapore@partridgepublishing.com

www.partridgepublishing.com/singapore

# CONTENTS

# CHAPTER 1

# UNIVERSE

## Genesis

The universe "began" with a big explosion, the so-called Big Bang, in which all celestial bodies were flung out from the centre of the universe. If Einstein's $E = mc^2$ turns out to be wrong, what caused the Big Bang? Why do all celestial bodies possess tremendous kinetic energy that has enabled them to propel the expansion of the entire universe, even until the present time?

As the universe ages, it continues to expand, with the gravitational attraction among all celestial bodies getting weaker and weaker. At the same time, more and more monomegastars transform into black holes after they run out of nuclear fuel. Lack of fuel eventually leads to subsiding heat intensity, and thus a failure to keep all atoms at arm's length from one another, especially in the cores of the monomegastars.

Despite the universe's continuing expansion, those dying monomegastars consolidate their mass further by transforming into black holes, exerting strong mutual attraction on nearby stars and pulling them closer. The nearby stars are consumed by the black holes, adding to the enlarging superstructure. The black holes become ever more massive.

Basically, black holes are superstructures made out of preons. When dying, gigantic stars collapse under their own weight, their atoms, especially in the core, are crushed into quarks and reduce further to preons. Positively charged preons are surrounded by negatively charged preons, and vice versa, to form a superstructure.

Unfortunately, only a handful of quantum scientists believe that a black hole consists of preons. This is because we can detect the presence of quarks only in a hadron collider, after collisions among hadrons. Therefore, preons are still pretty much a theoretical concept.

Since all negative quarks and leptons can interact with photons, we are certain that negative quarks and leptons are not the finest forms of particle. We suspect they contain some positively charged preons. Photons adhere to them because the photon is a negatively charged particle.

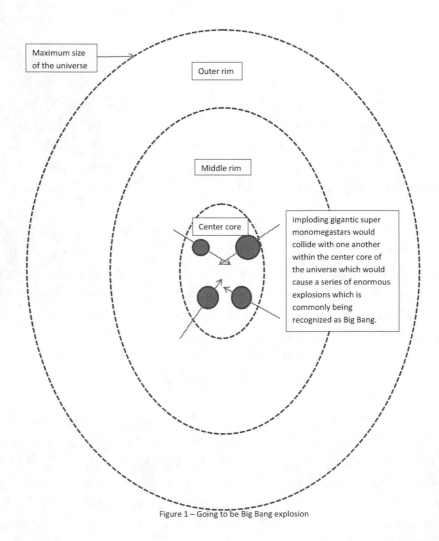

Maximum size of the universe

Outer rim

Middle rim

Center core

Imploding gigantic super monomegastars would collide with one another within the center core of the universe which would cause a series of enormous explosions which is commonly being recognized as Big Bang.

Figure 1 – Going to be Big Bang explosion

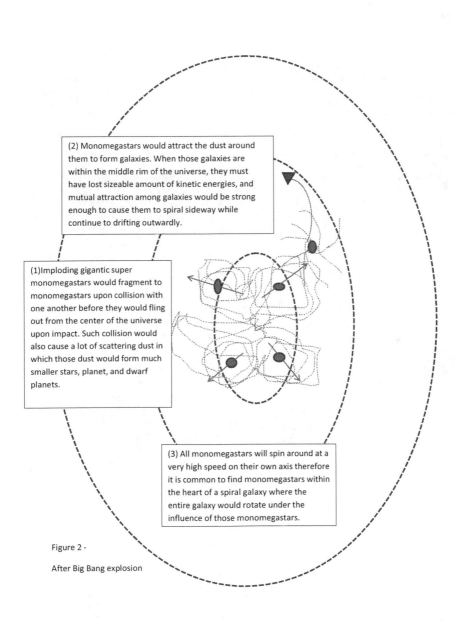

(2) Monomegastars would attract the dust around them to form galaxies. When those galaxies are within the middle rim of the universe, they must have lost sizeable amount of kinetic energies, and mutual attraction among galaxies would be strong enough to cause them to spiral sideway while continue to drifting outwardly.

(1)Imploding gigantic super monomegastars would fragment to monomegastars upon collision with one another before they would fling out from the center of the universe upon impact. Such collision would also cause a lot of scattering dust in which those dust would form much smaller stars, planet, and dwarf planets.

(3) All monomegastars will spin around at a very high speed on their own axis therefore it is common to find monomegastars within the heart of a spiral galaxy where the entire galaxy would rotate under the influence of those monomegastars.

Figure 2 -

After Big Bang explosion

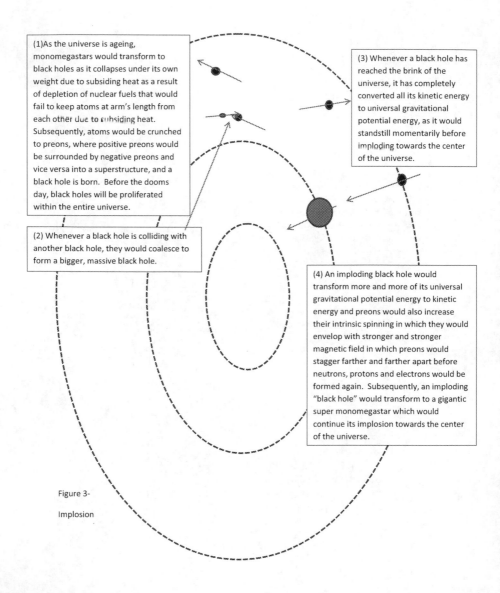

(1)As the universe is ageing, monomegastars would transform to black holes as it collapses under its own weight due to subsiding heat as a result of depletion of nuclear fuels that would fail to keep atoms at arm's length from each other due to subsiding heat. Subsequently, atoms would be crunched to preons, where positive preons would be surrounded by negative preons and vice versa into a superstructure, and a black hole is born. Before the dooms day, black holes will be proliferated within the entire universe.

(2) Whenever a black hole is colliding with another black hole, they would coalesce to form a bigger, massive black hole.

(3) Whenever a black hole has reached the brink of the universe, it has completely converted all its kinetic energy to universal gravitational potential energy, as it would standstill momentarily before imploding towards the center of the universe.

(4) An imploding black hole would transform more and more of its universal gravitational potential energy to kinetic energy and preons would also increase their intrinsic spinning in which they would envelop with stronger and stronger magnetic field in which preons would stagger farther and farther apart before neutrons, protons and electrons would be formed again. Subsequently, an imploding "black hole" would transform to a gigantic super monomegastar which would continue its implosion towards the center of the universe.

Figure 3-

Implosion

Before a dying, massive star transforms into a black hole, excess stationary photons are excreted from the newly formed superstructure in an explosive manner. This triggers a gigantic explosion that is recognized as a supernova.

Of course, a more massive dying star has greater mass and crunches harder than a lighter dying star. A more massive dying star also possesses a larger stockpile of stationary photons than a lighter dying star. Therefore, a more massive dying star turns into a brighter supernova than a lighter dying star.

The superstructure of black holes is very sturdy. When two black holes collide, they coalesce into one, more massive black hole. So far, according to astronomical observations, none of these black holes ever turns into a white hole. This author believes that black holes remain what they are, even after the universe has expanded to its maximum size before implosion takes place.

As the universe ages, as it expands further, more black holes emerge. This increases the gravitational clout among black holes and facilitates the implosion process, which will take place in the future as the universe expands at ever-slowing rates. In other words, remnants of initial kinetic energy will work against the gravitational forces, with less and less kinetic energy being converted to gravitational potential energy. Therefore, we believe all galaxies will slow down as the universe expands.

We can see the past of a segment of the universe, because the slowing of the universe's expansion allows light from the past to catch up with us now. Celestial bodies were moving much faster than the speed of light during the Big Bang and shortly afterwards, as the universe expanded at a neck-breaking pace. But this expansion has slowed. Matter, such as our earth, is now moving far more slowly than light. This allows the light of the past universe to catch up with us. We can assess it through astronomical observations.

Only after the universe has expanded to its maximum size will all celestial bodies at the brink of the universe have converted all their kinetic energy to universal gravitational potential energy. They will stop momentarily before they implode towards the centre of the universe, under the tugging of mutual gravitational forces.

All black holes and other celestial bodies gradually increase their velocities as they drift farther away from the brink of the universe towards the centre, by converting more of their universal gravitational potential energy back to kinetic energy.

Preons in the superstructure of a black hole will strengthen their magnetic energy after increasing their intrinsic spin. Preon spin will increase due to conversion of universal gravitational potential energy as the black hole moves away from the brink of the universe. Imploding black holes, to a certain extent, enable the preons in their superstructures to continue to spread apart. The strengthening of their magnetic energy disintegrates the entire superstructure. All negative preons become free from the grip of positively charged preons and vice versa.

This allows the superstructure to transform from preons to quarks again. Subsequently, some negative quarks will transform into electrons that circulate around a nucleus. Some will bind with positively charged quarks to form neutrons. Other positive quarks will transform into protons. The drifting superstructure of the former black hole will disintegrate to allow the formation of massive atoms during implosion.

Intense magnetic fields of electrons, neutrons, and protons allow them to amass more stationary photons if they ever make contact with photons. Freshly produced atoms are hungry for stationary photons. So the angular momentum of the electrons, neutrons, and protons in those atoms is very strong. Energetic neutrons and protons allow the formation of nuclei of very dense elements. The angular momentum of nucleons that results from saturation with stationary photons enables them to overcome the presence of more protons within the nucleus, facilitating the formation of very dense atoms.

In short, all celestial bodies in the universe will eventually implode towards the centre of the universe. Once they implode, they will hardly change their course, provided they are colliding with one another. All celestial bodies will increase their velocities as they inch towards the centre of the universe. Their speeds could be several times faster than the speed of light. When those fast-moving celestial bodies collide near the centre of the universe, the scene is beyond our imagination.

Since not all imploding celestial bodies would reach the centre of the universe at the same time, a Big Bang event could last for quite a while. The time lapse between the first and last implosions would be rather long. When the first, fast-moving imploding celestial bodies collide near the centre of the universe, they will fling matter out violently in an event recognized as a

Big Bang. Thus, the Big Bang is definitely not a single, gigantic explosion, as widely believed.

Initially, all celestial bodies drift radially from the centre of the universe after their collisions, due to their enormous kinetic energy. They fragment into smaller celestial bodies that are flung from the centre. The smaller bodies cluster to form galaxies as they move farther away from the centre. They transform their kinetic energy to universal gravitational potential energy as they move, allowing mutual gravitational attraction. This causes galaxies to drift "sideways", as if in an outward spiral. Of course, any such description of their motion is relative, not absolute.

This author believes that every monomegastar that emerged after the Big Bang is a big fragment of a former black hole. Therefore, monomegastars are humongous stars that are found at the heart of every gigantic galaxy. Monomegastars cause all celestial bodies in a galaxy to gyrate in tune with their movements. This indicates how massive monomegastars are. Monomegastars are likely candidates to transform back into black holes when their nuclear fuel has been completely depleted.

The dust that formed as a result of collisions among celestial bodies during the Big Bang coalesced to form star systems like our solar system. Bigger clusters of dust consolidated to form small stars like our sun, and fine dust consolidated to form planets. Very refined dust formed dwarf planets like the ones in the Kuiper belt.

The universe is in a late stage when its black holes and other celestial bodies have transformed all their kinetic energy to universal gravitational potential energy. At that point, the universe has inflated to its maximum size. Black holes and other celestial bodies will then stand still momentarily before imploding towards the centre of the universe to initiate the next round of the Big Bang cycle.

The universe has neither beginning nor end, in keeping with the laws of conservation of energy and conservation of matter. Matter cannot be created nor destroyed; the universe is finite in terms of its mass and energy. Midway through the implosion, most black holes will disintegrate to form super monomegastars, which will fragment into monomegastars and dust after colliding during the Big Bang.

In short, the Big Bang signifies the beginning of an expansion of the universe. The universe continues to expand until it reaches a maximum size determined by the complete transformation of kinetic energy to universal gravitational potential energy. At that point, implosion takes place. When imploding celestial bodies meet near the centre of the universe, the Big Bang takes place again. This dynamic cycle repeats endlessly.

## Newton's Universal Gravitational Law May Be Flawed

Newton realized that among matter, there exists a mutual attractive force between two objects. This attraction is called *gravitational force*. Gravitational force $F$ is equal to $G$, the universal gravitational constant, multiplied by the mass of the first object, multiplied by the mass of the second object, divided by the square of the separated distance between these two objects. The equation for this relationship is:

$$F = Gm_1m_2/r^2$$

Cavendish determined the value of $G$ with an apparatus consisting of two small spheres, each of mass $m$, fixed to the ends of a light horizontal rod suspended by a fine fibre. Two larger spheres, each of mass $M$, were placed near the smaller spheres. The attractive force between the smaller and larger spheres caused the rod to rotate. The angle at which the suspended rod rotated was measured by the deflection of a light beam reflected from a mirror attached to the vertical suspension. The experiment was carefully repeated with different masses of $M$ and $m$ at various separations.

In addition to providing a value for $G$, the results of Cavendish's experiment showed that the force is attractive, proportional to the product of $mM$, and inversely proportional to the square of the distance $r$.

The experiments carried out by Cavendish were not without flaws. The initial condition of the spheres was static. In addition, the earth exerted gravitational attraction upon all the spheres, especially the smaller spheres that were pinned. This made it harder for the horizontal rod to rotate freely.

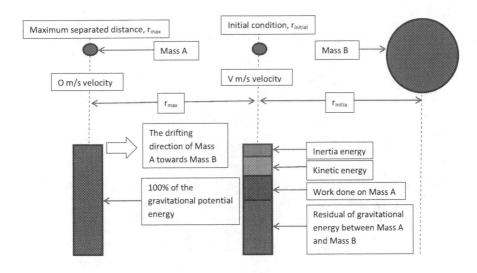

Figure 4 – Energies that locked between Mass A and Mass B

Let's assume that there are two spheres, one small (Mass A) and one big (Mass B), in the entire universe. We can pull the smaller sphere away from the bigger one to a distance $r_{max}$, representing the maximum distance at which the smaller sphere will start to drift towards the bigger sphere if we release our grip on it. The gravitational energy between the smaller sphere and the bigger one is at its lowest effective level at this distance.

If we let go the grip on Mass A, it will drift towards Mass B, moving faster as it gets closer. This increased speed occurs because as they get closer, the mutual attractive force between them gets stronger.

Cavendish's experiment eliminated the inertia energy, or the work done by Mass A to move from the theoretical maximum distance $r_{max}$ to the actual initial distance $r_{initial}$ at the beginning of the experiment, as shown in Figure 4. We need to find a way to estimate this inertia energy and to work out Mass A's total lost kinetic energy. We also need to calculate the work done by Mass A to move from its maximum distance to the initial condition of Cavendish's experiment.

Only after including the inertia energy, lost kinetic energy, and work done by Mass A can we enhance the accuracy of Cavendish's experimental results.

Since Cavendish's experiment was defective and inaccurate, any prediction of the amount of dark matter in the universe that is based on his computations may not be accurate. Newton's universal gravitational law is defective in this manner. Therefore it fails to help us understand the current state of the dynamics of the universe in real time.

The genesis of unification theory tells us where the kinetic energy of those celestial bodies originates and also predicts their tremendous velocities during the Big Bang. Apparently the prediction of the existence of dark matter has turned out to be untrue.

## Why Do Life Forms Exist in the Universe?

Generally, the existence of life forms in the universe is due to the dynamic nature of the universe itself. In other words, the dynamics of the universe allow the transformation of one form of energy into other types of energy. An exchange of photons can take place. This leads to the emergence of life forms.

During the Big Bang, the core of the universe became extremely hot and volatile. Kinetic energy was at its peak. Almost the entire stockpile of the universe's dynamic photons saturated the centre. Therefore, no life forms existed during or shortly after the Big Bang.

After the Big Bang, most celestial bodies that had strong kinetic energy drifted away from the core of the universe. Remnant bodies with low kinetic energy were trapped in the core. Even so, the angular momentum of the nucleons of their atoms could be considered somewhat strong, and was enveloped in a strong magnetic field too. The nucleons had an uncanny appetite for stationary photons and exchanged photons with their surroundings. As a result, the universe's core rapidly became cool long and could hardly allow the existence of any life forms.

When habitable planets drift in the middle portion of the universe, we reckon that nearly half of their total energy is kept in the form of universal gravitational potential energy. Over time, less of their kinetic energy is converted to universal gravitational potential energy, because by this stage of the expansion of the universe, such a planet must have slowed down significantly already. Conditions in the middle portion are moderate and conducive to nurturing and sustaining life forms because chemical reactions can take place in a sustainable, orderly manner. This buys time for life forms to thrive for a long time before the planet progresses to the outer rim of the universe.

Even as less kinetic energy is converted to universal gravitational potential energy, the universe continues its moderate rate of expansion. The chemical environment of a habitable planet shifts gradually in a dynamic equilibrium that allows life forms to thrive. In other words, the physical and chemical characteristics of elements and their compounds as we now know them are not static, but subject to change in parallel to the gradual expansion of the universe.

The DNA of a cell contains three distinctive pieces of information: the anatomy map of the cell (for multiple-cell structures, it also contains the mapping of its position relative to other cells), the anatomical structure of the cell, and all chemical processes that take place within the cell (such that active segments of DNA synthesize chemical compounds to facilitate the functioning of the cell).

The anatomy map ensures that the different components of the cell are pieced together correctly to achieve the cell's intended function. The information about the anatomical structure of the cell details the construction of each individual component. The information about the chemical processes of the cell control the synthesis of chemical compounds to facilitate the functioning of the cell. The cell's function and its DNA must be mutually compatible in terms of anatomical positioning, anatomical structure, and chemical processes. Cancerous cells, for example, cannot repair themselves because their wrecked DNA is no longer compatible with the cell's intended function.

The emergence of a first cell definitely does not happen by chance. Cell function and cell DNA must always be mutually compatible. Similarly, a collection of cells in which the DNA of individual cells may not share the same information would fail to work together to become a new creature.

A complex creature must have complex enough DNA to provide anatomical mapping, anatomical structure, and chemical processes. For instance, a particular cone cell in a human retina has DNA information regarding its position relative to the rest of the cone cells that is necessary to create the conditions for vision. Likewise, the cell has information regarding its anatomical structure that must match the anatomical structure of other cone cells. Lastly, the cone cell must capable of synthesizing chemical compounds that precisely facilitate the functioning of the cone cell, on its own and in concert with other cone cells. Every one of our cells has identical DNA, but only the relevant segment for that particular cell can be "on" so that its positioning is correct, structurally sound, and chemically relevant.

Darwin's theory of evolution may be wrong. One species might not evolve into another species even if their DNA is compatible, like the sterile mule which has a horse and a donkey as parents. Without doubt, the DNA of every species is constantly forced to adapt to alterations in the environment. There are constant modifications to the physical and chemical characteristics of the chemical components of DNA, such as N-containing bases, as a result of the continuous expansion of the universe. The DNA of a palaeoman is different from our DNA in term of its structure and composition. As a result, the appearance of a palaeoman is different from the appearance of a modern man.

This notion strongly suggests that humankind did not evolve from an ape. Our existence on earth did not happen by chance, because all of us have the same pairs of chromosomes and number of genes. One more or one less chromosome or gene results in congenital disease. The DNA in every cell of an individual must be identical to one another. These facts likewise indicate that the first people on earth did not appear by chance. Therefore it is only fair to embrace the idea that the life forms in the universe are God's creation.

Even if, one day, humans manage to construct a cell artificially, it will not turn into a life form unless God blows life into it. This is obvious if we look at non-living things, which are dead because they have no souls inside. We are alive because our souls are still strapped within our bodies.

We are inevitably forced to harmonize with our environment, since we are what we eat. This warns us of the possible hidden dangers of consuming GM foods. We still have little knowledge about their safety and impact on us. What if we contract gastrointestinal cancers after consuming GM foods? Worse still, what if several generations pass before we come to realize that GM foods cannot harmonize with us? By then, the deadly, irreversible impact of GM foods could spell the end of our existence on earth. Why don't we allow ourselves to harmonize with natural food rather than play God blindly?

All life forms will cease to exist when a formerly habitable planet enters the outer rim of the universe, where almost all of its kinetic energy has been converted to universal gravitational potential energy. Matter hungers for stationary photons when its universal gravitational potential energy has reached its maximum, which is one of many special characteristics of matter. This dormant condition is hostile to life forms. Stars in outer rim will be much dimmer, mainly because they have almost run out of nuclear fuel. They too will have an increased appetite for stationary photons.

The extremely cold environment hinders the existence of life forms at the centre of the universe long after the Big Bang. Similarly, conditions are hostile to life forms at the outer rim of the universe. Life forms thrive when habitable planets, like the earth, are in the middle portion of the universe, because moderate conditions are conducive to the existence of life forms. Life forms exist not only on earth, but everywhere in the middle portion of the universe.

# CHAPTER 2

# ASTRONOMY

## Stars – Fission or Fusion?

Stars emit enormous amounts of light and heat. The sun, our closest star, provides us with heat and light during the day so that life can be sustained on earth. Why and how do stars generate heat and light?

We know with certainty that Einstein's $E = mc^2$ is not true. Photons are particles that possess mass. During the process of fission, stationary photons that saturate nucleons are dissipated to the atom's surroundings.

A fusion process is unrealistic, since to compress more nucleons to make a denser nucleus structure would require packing more stationary photons in order to strengthen angular momentum. Only more angular momentum could resist the greater repulsive forces of an increased number of protons within the nucleus. Fusion might have taken place during the Big Bang, when nucleons had their maximum amount of kinetic energy and were surrounded with very strong magnetic energy. In addition, an abundance of dynamic photons in the centre of the universe makes it easier to pack dynamic photons onto those nucleons. With more angular momentum, more nucleons could cluster together to form a stable, dense, unique nucleus structure.

It is obvious that stationary electrons possess kinetic energy, since they are confined within an atom and constantly revolving around the nucleus at approximately 80 per cent of the speed of light. Nucleons also possess kinetic energy, but this is not obvious. The nucleons of a solid object stabilize themselves within the nucleus structure as if they were stationary, while thrusting in and out of the nucleus structure.

Nucleons possess kinetic energy, but they keep it in a different form: angular momentum, which is a force that acts on the nucleon as shown in Figure 5. When nucleons cluster together to form the unique nucleus structure of a stable element (Figure 6), the electrostatic force, magnetic force, gravitational force, and angular momentum of all nucleons must be stabilized.

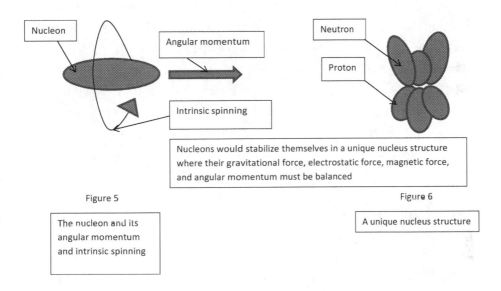

Figure 5

The nucleon and its angular momentum and intrinsic spinning

Figure 6

A unique nucleus structure

There is a close association between intrinsic spin and magnetic energy. Tapping angular momentum hastens intrinsic spin, fostering a stronger magnetic field around the nucleon. Likewise, weakening magnetic energy and intrinsic spin boosts angular momentum.

The presence of stationary photons on nucleons boosts their magnetic energy and angular momentum. Therefore, nucleons that are saturated with more stationary photons possess stronger angular momentum as well as greater magnetic strength. Releasing stationary photons to the surroundings weakens the nucleon's angular momentum and magnetic field strength.

During the Big Bang, an abundance of dense radioactive substances were produced. Of course, there are a lot of different types of dense radioactive substances. One of them is uranium, which is rather common on earth. We don't know with certainty if there are other substances that are much denser than uranium in the core of the earth. Uranium is the densest natural radioactive substance found in the earth's crust.

The nuclei of very dense radioactive substances have more nucleons compressed in them, and the nucleons are saturated with many more stationary photons than those of ordinarily dense radioactive substances. Very dense radioactive substances release more dynamic photons to the surroundings when their nuclei undergo radioactivity.

Very dense radioactive substances have much shorter half-lives than dense radioactive substances. Our sun has surely outlasted have experienced supernova explosions. Our sun is not as volatile; its core is filled with dense radioactive substances, as compared to the cores of gigantic stars, which were filled with very dense radioactive substances prior to explosion.

Stars generate heat and light through fission of their nuclear fuel. Star cores are filled with dense radioactive substances, very dense radioactive substances, or both. Generally, very dense radioactive substances decay at a much faster rate than dense radioactive substances.

## Main Sequence

So all stars in the main sequence are not too big, and their cores are mainly filled with dense radioactive substances. Therefore, all young stars in the main sequence will have moderate luminosities. As they age, a big portion of their dense radioactive substances decay, and their heat and light progressively dissipate. The reduced heat no longer keep atoms in the core apart. The atoms crush together under their own weight as the star dims.

Since the weight of this kind of dying star is not massive, the crush is not intense enough to become an implosion. The dying star turns into a red dwarf. When all its radioactive substances have been depleted, they simply cease to emit light and heat.

## Hypergiant Stars

All hypergiant stars are extremely bright (about a million times brighter than the sun) due to their enormous size. Their cores may be filled with dense radioactive substances, very dense radioactive substances, or both. Since they emit tremendous heat and light, their high-temperature environment make the already unstable nucleus structure weaken further. Therefore both dense and very dense radioactive substances decay at a faster pace in hypergiants than in dwarf stars. As a hypergiant ages and collapses, the tremendous crushing of its core atoms reduces them to quarks and then to preons.

The production of preons dissipates a large volume of dynamic photons to the surroundings. Immediately those preons form a very sturdy superstructure, in which positive preons are surrounded by negative preons and vice versa. Soon the superstructure grows in density. Tremendous heat dissipates from the core. This leads to a gigantic explosion, commonly known as a supernova.

The preons conserve their angular momentum and intrinsic spin, each revolving on its own axis at high speed. This allows the preons to be surrounded with enormous magnetic fields.

The entire superstructure rotates at neck-breaking speed because the preons are conserving their angular momentum and kinetic energy. A black hole is born, as shown in Figure 7.

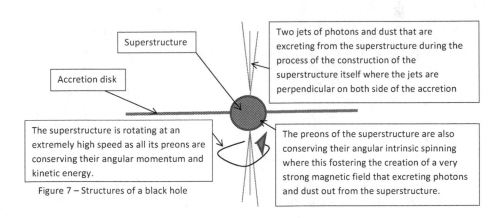

Figure 7 – Structures of a black hole

Since the superstructure is enormously dense and rotating at high speed, it attracts the dust and stars around it to form an accretion disk within its enormous gravitational pull. The jets of dynamic photons and "dust" are perpendicular to the sides of the accretion disk. Therefore we strongly believe that there are two equal magnetic fields that are symmetrical along the accretion disk, as shown in Figure 8.

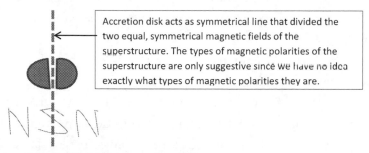

Accretion disk acts as symmetrical line that divided the two equal, symmetrical magnetic fields of the superstructure. The types of magnetic polarities of the superstructure are only suggestive since we have no idea exactly what types of magnetic polarities they are.

Figure 8: A black hole's magnetic polarities

Any debris and stars dragged into the accretion disk spin closer and closer to the superstructure. Eventually, they plunge into the superstructure. The enormous gravitational forces of the superstructure rips those atoms apart before reducing them to quarks to fuel further construction of the superstructure. Excess photons and dust are excreted from the superstructure through magnetic-magnetic interactions. The dust is composed of diamagnetic objects that could be as big as boulders or bigger.

A black hole grows more massive as it is fed with stars and other celestial bodies. Normally the emergence of a black hole is a sign that the universe is starting to age. The function of a black hole is to consolidate mass in the universe even as the universe continues to expand, making the next round of implosion possible.

A black hole will only disintegrate during an implosion, when universal gravitational potential energy has been transformed to kinetic energy. A stronger magnetic field than the one enveloping the preons eventually rips the superstructure apart, setting the stage for the next round of the Big Bang.

All hypergiant stars transform to black holes before they run out of nuclear fuel. Their enormous mass ensures that atoms are crushed much harder when a hypergiant is collapsing under its own weight.

## Supergiant Stars

Supergiant stars have a luminosity ten thousand times that of the sun. Their cores may be filled with dense radioactive substances, very dense radioactive substances, or both. Supergiants filled with very dense radioactive substances are less massive than those filled with dense radioactive substances, since very dense radioactive substances release a greater intensity of dynamic photons during radioactivity than do dense radioactive substances.

When supergiant stars age and collapse, a process similar to the formation of hypergiant black holes takes place. Dissipating heat from the core results in a supernova explosion. Since more massive, dense radioactive substances crush better than less massive, very dense radioactive substances, supergiants composed of dense radioactive substances will in some cases become black holes. Because they are less massive, supergiant stars consisting of very dense radioactive substances collapse into less pure superstructures because the crush of atoms is not as efficient. these less pure superstructures result in so-called *neutron stars*.

It would be naive to think that a neutron star consists only of neutrons. Neutrons need protons to preventing them from decaying. In turn, neutrons provide interlocking to protons, making the atomic nucleus cohesive. Electrons of the neutron are attracted to the positive quarks of protons. This keeps the unique nucleus structure cohesive while maintaining the stability of the neutrons. Protons need neutrons to defuse the repulsive forces between them. Without protons, a cluster of neutrons would soon disintegrate into free electrons, protons, and quarks. There is no doubt that a neutron star is very massive, but it is not made solely out of neutrons. It consists of quarks instead.

Very massive supergiant stars turn into black holes. Less massive supergiant stars turn into neutron stars.

## White Dwarfs

A white dwarf is approximately one one-hundredth of the brightness of the sun and much less massive. But most white dwarfs are likely to end in a type Ia supernova explosion. How and why can a less massive star undergo this intense explosion?

When a white dwarf undergoes a type Ia supernova explosion, many new, bright, smaller stars are born in the dust. A white dwarf core is filled with very dense radioactive substances.

As a white dwarf ages, it releases less heat and fails to keep its core atoms separated, collapsing under its own weight. The compression, as with hypergiant and supergiant collapses, causes an explosion. But this explosion is much less intense than a true supernova, and the lesser crush of the atoms does not produce anything close to a superstructure.

Dust is widespread in the vicinity of an exploded white dwarf. The accumulation of remnants of very dense radioactive substances eventually forms smaller, bright, newborn stars that emit heat and light for a short duration. The exploded white dwarf itself cools down further before it ceases to emit heat and light.

Though white dwarfs are rather small, they consist of very dense radioactive substances that are decaying at very fast rate. This makes white dwarfs appear rather bright. Aging white dwarfs undergo Ia supernova explosions before ceasing to emit light and heat.

## Neutron Stars and Black Holes

Research needs to be carried out to see if all neutron stars eventually convert to black holes. Black holes and neutron stars share a similar structure. Both have fast-rotating accretion disks. Both have two jets of photons and dust that spurt perpendicularly from the superstructure on each side of the accretion disk. Therefore, both black holes and neutron stars must have superstructures embedded in the centre of their accretion disks.

Our assumption is that the superstructure of a neutron star is not pure—that is, it is composed of quarks, which are not fully decayed, fundamental particles. This is in contrast to the superstructure of a black hole, which is composed of preons. Can the impure superstructure of a neutron star be amended to become pure after it has gained more mass by gobbling up other celestial bodies?

We reckon that the answer is yes. A neutron star will eventually transform into a black hole after it has swallowed enough mass. The tremendous weight will manage to crush quarks into preons to fuel the construction of a pure superstructure.

## Stars – Dense and Very Dense Radioactive Substances

Stars are generally huge celestial bodies. Their cores are filled with dense radioactive substances, very dense radioactive substances, or both. Due to the expansion of the universe, kinetic energy is transformed to universal gravitational potential energy as the stars drifting from the centre. The transformation weakens the angular momentum of nucleons, nudging restructuring of the unique nucleus structure.

Nucleons of dense substances are saturated with stationary photons. Somehow the weakening nucleons trigger restructuring of the nucleus. Simpler, smaller nuclear structures require less angular momentum to maintain stability. Therefore, excess stationary photons are released to the surroundings as dissipated heat during radioactivity.

Very dense radioactive substances likely undergo alpha emissions. Dense radioactive substances may also undergo alpha emission, though rarely, during which helium nuclei are released to the surroundings as the nucleus splits into several smaller daughter nuclei. The helium nuclei steal electrons from somewhere else before emerging as helium atoms.

Dense radioactive substances always undergo proton emission, neutron emission, or both. Very dense radioactive substances sometime also undergo proton emission or neutron emission. During proton emission, a single proton steals some electrons from its surroundings and transforms to a hydrogen atom. During neutron emission, a single neutron that has not been absorbed by a nucleus quickly decays to produce a single proton, some quarks, electrons, and heat. Such a single proton eventually turns into a hydrogen atom by stealing electrons from somewhere else.

All spectra of type O stars with brightest luminosity show the prevalence of hydrogen and helium gases in their atmospheres. This clearly demonstrates that all stars undergo fission to release heat and light, and that most stars are made out of a mixture of dense and very dense radioactive substances.

# Non-Radioactive Hot Stars Preceded Radioactive Stars

What type of stars came before type O stars? Only hydrogen and helium fingerprints are present in the spectra of type O stars, which indirectly proves that radioactive decay of dense and very dense radioactive substances has already started. Stars that are saturated with dense and very dense hot, stable elements are candidates for the type of stars that precede type O stars.

All stars, big or small, created during the Big Bang are saturated with enormous numbers of stationary photons. Some of those elements are much denser than uranium. Others are just as dense. As those stars drifting away from the centre of the universe, more of their internal energy, especially angular momentum, is weakened, gradually transforming to universal gravitational potential energy. This causes them to release dynamic photons to the surroundings while they still remain as stable elements. Therefore, "non-", or type N radioactive hot stars precede type O stars.

The continuous expansion of the universe further weakens the angular momentum of type N stars, to the extent that they are unable to resist the repulsive forces among the high number of protons within their unique nucleus structure. Restructuring of that structure is inevitable. Smaller and simpler daughter nuclei replace the original nucleus. Formerly stable, hot elements turn into radioactive substances as type N stars transform to type O stars.

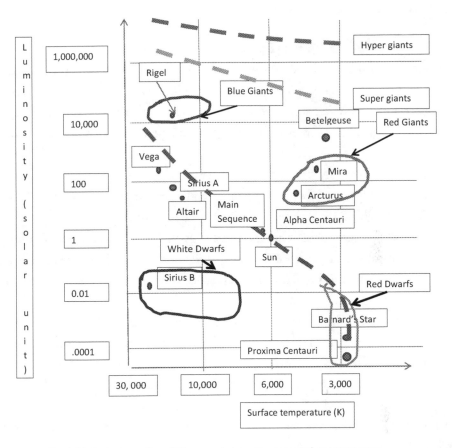

Graph 1 – Hertzsprung-Russell Diagram – Luminosity versus surface temperature

Type N stars release heat and light to their surroundings without undergoing any radioactivity. As they drift away from the centre of the universe, their stationary photons transform to dynamic photons, to be dissipated to the surroundings. This process further weakens the angular momentum of the star's nucleons in a way that hastens the transformation of what used to be non-radioactive substances into radioactive substances.

We don't know how bright type N stars were. It depends upon how fast they drifted from the centre of the universe. Of course, no one was present to witness how those first stars shaped the entire universe in the beginning.

# Classifying Stars Based on Spectra

The first hot stars are type N. Following type N are O, B, A, F, G, K, and M. Table 1 shows these types with their surface temperatures, prominent absorption lines, and familiar examples. Typing of stars is based on their spectra. Most stars progress from type O to type M as they age. Ultimately, they cease to emit light and heat when they run out of nuclear fuel.

It is not necessarily true for type F, G, K, and M stars that their respective hydrogen absorption lines are moderate, relatively faint, faint, and very faint. As stars age, the accumulated hydrogen gas in their atmospheres must be enormous. The release of other gases as a result of radioactive decay dilutes the concentration of both helium and hydrogen gases. Due to decreasing heat and light dissipating from these types of stars, the hydrogen absorption lines in their spectra may not be obvious.

For instance, the sun's spectrum as taken on earth show helium and hydrogen absorption lines that are very strong, contrary to the definition of a type G star, which we know the sun is. Table 1 indicates that the hydrogen absorption lines are relatively faint in type G stars.

The most striking feature in Graph 1, which categorizes all types of stars in term of luminosity versus surface temperature, is that Betelgeuse (M2) is very bright at approximately 8,000 solar units, but Barnard's Star (M5) measures 0.0009 solar unit. Both Betelgeuse and Barnard's Star are M type stars, but Betelgeuse is much brighter. This could be due to the enormous size of Betelgeuse.

The luminosity of a star depends upon its size and the radioactive substances in its core. Unfortunately, there is no way we can know this information with our current technology. All we have is *perceived* luminosity, which is always affected by how far away the star is from earth.

Let's say one star composed of dense radioactive substances and another, smaller star composed of very dense radioactive substances are at similar distances from the earth. The smaller one could be perceived to be much brighter than the bigger one simply because very dense radioactive substances dissipate more heat than dense radioactive substances. In addition, very dense radioactive substances have much shorter half-lives than dense radioactive substances.

Dense radioactive substances dissipate less intense heat. They sometimes undergo alpha emission, but they undergo proton and neutron emission most of the time as they decay. Type M stars have used up most of their very dense radioactive substances, leaving less dense radioactive substances in their cores that release less intense heat during decay. Therefore their spectra are rather dark.

This doesn't imply that all type M stars have hydrogen absorption lines that are very faint. On the contrary, their hydrogen absorption lines may still be fairly obvious, since hydrogen accumulates from the moment a star is born.

| Spectral Class | Surface Temperature (K) | Prominent Absorption Lines | Familiar Examples |
|---|---|---|---|
| N | Unknown | Clear lines in the spectrum | |
| O | 40,000-30,000 | Bright and dark lines | Zeta Puppis(O6) |
| B | Over 25,000-12,000 | Helium dominant | Rigel (B8) |
| A | 10,000-8,000 | Hydrogen strong | Vega(A0) Sirius(A1) |
| F | 7,500-6,000 | Calcium dominant; hydrogen moderate | Canopus(F0) |
| G | 6,000-5,000 | Metallic lines appear; hydrogen relatively faint | Sun(G2) Alpha Centauri(G2) |
| K | 4,000-3,000 | Strong metallic lines; hydrogen faint | Arctururs(K2) Adebaran(K5) |
| M | 3,000 | Neutral atoms strong; a lot of presence of other less dense and dense elements; bands due to molecules; hydrogen very faint | Betelgeuse(M2) Barnard"s Star(M5) |

Table 1: Various spectral classes of stars

Likewise, it is not true that helium absorption lines vanish completely from the spectra of M type stars. Helium is still prominent because it has been accumulating since the star was born. As the star ages, the release of other gases, such as nitrogen and carbon dioxide, and the eighty-eight lightest elements and their compounds dilute the helium and hydrogen gases in the star's atmosphere. In addition, the less intense heat and light that dissipate from type M stars make hydrogen and helium absorption lines less obvious, especially in smaller type M stars.

The dotted dark red line in Graph 1 is called the *main sequence*. Most of the stars in the universe are believed to fall into this category. Most stars in the main sequence are believed to be rather bright (dwarfs) after birth but gradually become dimmer, like the sun. Eventually they become red dwarfs before ceasing to emit light and heat completely as dead celestial bodies.

Generally, stars are very hot shortly after their births, but gradually lower their temperatures. Hot, non-radioactive substances become radioactive substances. Most very dense radioactive substances are used up first, before dense radioactive substances. Since very dense radioactive substances are less stable, they have shorter half-lives and release more heat than dense radioactive substances.

Most stars evolve from N through O, B, A, F, G, and K to type M stars. They cease to emit light and heat when all the radioactive substances in their cores have been completely depleted.

Hypergiants, supergiants, and white dwarfs, as they cool, collapse under their own weights. This hastens the radioactive decay of their remaining very dense radioactive substances. Tremendous heat is released, resulting in a supernova. Hypergiants turn into black holes after going supernova. Supergiants may become either black holes or neutron stars, depending on their mass just before the explosion. White dwarfs may turn into something like red dwarfs after a similar but less intense Ia supernova explosion.

## Sun

The sun is the earth's closest star. It provides us with heat and light so that life can be sustained on earth. Its core is filled with mostly dense radioactive substances.

It is not true that the sun generates heat through fusion. Four protons are squeezed to form a helium nucleus under its tremendous gravitational force. Subsequently, a tremendous amount of heat is released. The amount of dissipated heat corresponds to Einstein's $E = mc^2$.

In earlier segment, it was shown that Einstein's $E = mc^2$ is not correct. It is not true that a minuscule amount of matter can transform into a tremendous amount of energy, because the photon is a particle that has mass. In other words, you must have photons to emit photons.

In addition, to construct a larger nucleus with more nucleons, especially protons, requires compacting those nucleons with more stationary photons to bolster their angular momentum. More angular momentum is needed to resist the stronger repulsive forces among a greater number of protons. Fusion requires the input of more energy and additional photons to bolster the angular momentum of those nucleons. It is impossible that energy could be tapped from fusion.

The core of the sun is filled with dense radioactive substances. The half-lives of those substances is long when compared to the half-lives of very dense radioactive substances. Therefore we can expect the sun to outlast much bigger stars filled with very dense radioactive substances, many of which have already undergone supernova explosions.

It is believed that the sun used to have a sizeable amount of very dense radioactive substances in its core. This led to an abundance of helium gas in the sun's atmosphere, based on the assumption that very dense radioactive substances undergo alpha emission more frequently. Emitted helium nuclei steal some electrons from their vicinity before turning into helium atoms. Sometimes very dense radioactive substances may undergo proton or neutron emissions too. Dense radioactive substances undergo proton and neutron emission most of the time. A single proton transforms to a hydrogen atom by stealing electrons from somewhere, and a free neutron converts to a single proton, electrons, and quarks. The single proton then converts to a hydrogen atom.

Most stars have the fingerprints of hydrogen and helium in their spectra. Helium and hydrogen are the by-products of radioactivity, which implies that most stars used to be enriched with a mixture of dense and very dense radioactive substances. There is normally an abundance of hydrogen and helium gases in all stars' atmospheres. It is a fallacy that all of them generate heat and light by combining four hydrogen atoms into a helium atom. Even the spectra of type M stars, like the one in Betelgeuse, clearly show the presence of hydrogen in their atmospheres. Since most type M stars, especially dwarfs, tend to release less intense heat, their spectra may not be bright enough to make the fingerprint of hydrogen noticeable. The detection of hydrogen in type M stars clearly proves that fusion is not feasible in stars.

Though the sun is categorized as a type G star, its spectrum show clear absorption lines for both hydrogen and helium gases in its atmosphere.

## Sunspots

Sunspots are large, dark patches on the sun. Because they appear dark, they are considered cooler. Sunspots normally are born near the solar north and south poles. They drift slowly towards the equator due to the sun's rotational movement. Normally, they burst into solar flares before they reach the equator. Why?

The sun is not homogeneous. It consists of various materials, namely ferromagnetic, diamagnetic, paramagnetic, and non-magnetic materials. Ferromagnetic materials become a stronger magnet when a stronger external magnetic field is present. Diamagnetic and paramagnetic materials turn into weak magnets under similar circumstances.

Magnetic fluxes are more saturated near the poles, therefore the magnetic fields near the solar poles are stronger. The poles are where sunspots are born. Due to convection within the mantle of the sun, a small chunk of ferromagnetic material drifting upward from (A) to (B) would merge with other chunks of ferromagnetic materials to form a much bigger chunk at (C) (see Figure 9). They would be attracted to one another due to their magnetic strengths.

The bigger chunk of ferromagnetic material cools as it drifts toward the surface. It also continues to attract more ferromagnetic material as its magnetism strengthens. Eventually it reaches the sun's crust, where it flattens under shearing forces from continuous convection movement. That makes it easier for the chunk to float on the sun's crust at (D) as it continues to drift slowly towards the equator.

A sunspot appears to be darker because, under the influence of an external magnetic field, ferromagnetic materials become stronger magnets, increasing their appetite for stationary photons. A sunspot appears darker because it is absorbing more photons from its surroundings. This doesn't imply that a sunspot is actually cooler.

Smaller chunks of ferromagnetic material near the sun's crust sink during convection at (E), forming a larger chunk of ferromagnetic materials at (F), as shown in Figure 9. Smaller chunks act like magnets attracted to other magnets, merging into a much bigger magnet. But as the larger chunk sinks, its temperatures gets very high. The intense heat shatters it intoto smaller chunks at (G). The new, smaller chunks flow upward at (B), entering another cycle of sunspot creation.

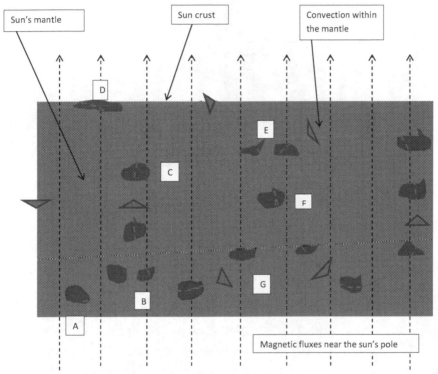

Figure 9: Formation of sunspots

The presence of stronger magnetic fields near the poles help sunspots to amass stationary photons in a domino effect. A sunspot that has amassed more stationary photons bolsters its magnetic strength, enveloping its ferromagnetic material in a stronger magnetic field that can hold even more stationary photons, and so on.

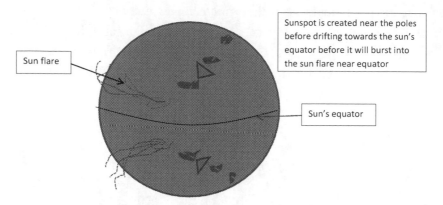

Figure 10: Sunspots and sun flares

Initially, a sunspot may grow in size as it drifts towards the equator. As it drifting gets farther from the pole, however, its distance from the pole's strong magnetic field reduces the sunspot's ability to retain its stationary photons. Suddenly, a sizeable number of stationary photons will be released to the surroundings as dynamic photons. This loss further weakens the sunspot's magnetic field, causing a further release of photons. This process can go on like a reverse domino effect.

Eventually, the abundance of dynamic photons being released causes a gigantic explosion within the sunspot, as shown in Figure 10. Ferromagnetic materials are flung into the air. This phenomenon is called a *solar flare*. The dynamic photons released in a flare assault any planets that happen to be facing the point of the flare.

The ferromagnetic materials that burst out during a solar flare eventually fall back into the sun. The dynamic photons, however, continue drift away.

A solar flare's assault on earth may cause power outages. The abundance of dynamic photons cuts through transmission cables, with some photons massing on free electrons in the cables. These electrons are likely to unload excess photons when they experience resistance during transmission. Excess heat is dissipated when free electrons pass through an automatic switch at a substation, causing the switch to trip and create a power outage. Similar massing and dissipation caused nineteenth-century telegraphy machines to tap incessantly during solar flares, even if the machines were offline.

Solar flares consist of clusters of dynamic photons that head out from the sun. It is not true that solar flares contain charged particles, such as single protons or dynamic electrons.

## Chromosphere and Corona

Astronomers believe that the sun's corona is much hotter than its chromosphere. Are they correct? Readings of the sun's spectrum indicate that temperatures in the corona exceed one million kelvins. Temperatures in the chromosphere are less than six thousands kelvins. These temperature estimates are based on the detection of iron that has been ionized in the corona versus iron ions in the chromosphere. Is the assumption that iron ions are an accurate proxy for temperature correct?

In the laboratory, specimens are subjected to the same localized gravitational potential energy. Let's say we are ionizing iron atoms. Each time we ionize them, the flexing of their unique nucleus structure is slightly different, because the iron atoms have fewer and fewer stationary electrons in their orbitals. Every time an atom is ionized, its unique nucleus structure adapts to the same localized gravitational potential energy at sea level. To ionize an iron atom at sea level requires a lot of energy.

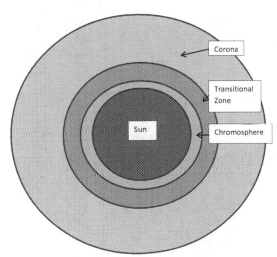

Figure 11: Corona and chromosphere

Similarly, we may find it challenging to ionize oxygen and nitrogen gases in the laboratory at sea level. But at altitude, oxygen and nitrogen molecules trapped in a storm cloud can be ionized with ease just by rubbing against other air molecules. We overlook the influence of localized gravitational potential energy in situations where, at a specific altitude, the unique nucleus structure of nitrogen and oxygen may flex in a different manner.

Ambient temperature at high altitude may also influence ionizing capability. Temperature may improve the atoms' capability to amass stationary photons from their surroundings, which would enable them to ionize rather easily.

Therefore, storm clouds can accumulate excess stationary electrons. Once a cloud starts to precipitate, icy crystalline structures are no longer present to help nitrogen and oxygen molecules to hold the electrons. The electrons are transformed into free electrons, and lightning ensues.

Maybe the altitudes in the solar corona are different from those in the chromosphere, in term of localized gravitational potential energy. This would cause atoms in corona to flex their unique nucleus structure differently, and thus be more easily ionized. In a laboratory setting, a comparatively large amount of energy is needed to ionize atoms, especially iron atoms. If we used laboratory data alone, we might conclude the corona must have a very high temperature that enables it to ionize iron atoms. But we would be wrong.

Figure 11 shows the corona sitting on top of a transitional zone. The transitional zone rests on the chromosphere. We know intense heat is emitted from the photosphere to the chromosphere. Atoms, especially nuclei that have absorbed excess stationary photons, experience increases kinetic energy in this heat. As they travel farther away from the photosphere, they may unload some of their stationary photons and slow down. If they continue to travel outward, they unload even more stationary photons and transforming even more kinetic energy to gravitational potential energy, slowing them further. Atoms cannot travel farther than the corona due to their low kinetic energy and the strong gravitational pull of the sun.

Any atoms that implode towards the core from the corona re-amass stationary photons, since the temperatures is higher at the centre. Atoms that collide with more energetic atoms in the chromosphere may be diverted back to the corona again. No doubt there are fewer atoms in the chromosphere, and those atoms possess extremely high velocities, or kinetic energy. Therefore their free paths are long. In addition, the dissipation of dynamic photons from the photosphere saturates the chromosphere, helping it to support the weights of the transitional zone and corona.

The corona is cooler, since it is located relatively far away from the sun's core. Atoms in the corona are compact, saturated, and cooler. There are slightly more atoms per volume per time present in the corona as compared to the chromosphere. As a result, more atoms exchange photons with their surroundings per volume per time. This creates the perception that in the corona, atoms undergo a more intense exchange of photons.

On the contrary: there are slightly fewer atoms per volume per time in the chromosphere. Certainly, the photosphere floods the chromosphere with a large quantity of dynamic photons per volume per time. Thus the atoms in the chromosphere undergo a very intense exchange of photons with their surroundings, despite the fact that they are travelling at tremendous velocity.

It is a fallacy to think that even at higher altitudes, atoms remain the same as they are at sea level. Higher localized gravitational potential energy allows the unique nucleus structure to flex in a way that greatly improves atoms' capability to amass stationary photons. At the same time, the greater potential re-orients atomic orbitals, making it easier for atoms to be ionized.

The slightly higher saturation of atoms in the corona versus the chromosphere affects our perception of photon exchange and creates the impression that the corona has a higher temperature than the chromosphere.

If we define temperature as the availability of a specific number of dynamic photons per volume per time, then it is impossible for a space that is much farther away from a heat source to have a higher temperature than a space that is closer to that heat source. The corona could not have a higher temperature than the chromosphere, which is located right above the photosphere.

It is important to understand the effect of localized gravitational potential energy on atoms at higher altitudes. This in turn betters our understanding of the real temperature distributions of the corona and chromosphere.

# Spinning-Gravitational Effects

All planets, including asteroids and dwarf planets, revolve around the sun in the same direction. Is this coincidence, or is this because of other prominent factors?

The sun also rotates on its own axis. The direction in which all planets revolve around the sun is the same as the sun's rotational direction on its axis.

Given a rather massive celestial body that is rotating in its own right, we can no longer assume the sun is a "point" mass or that all its mass can be replaced by a hypothetically equivalent point mass when we are analysing the movement of planets around it.

If it were not for the presence of the sun's spinning-gravitational effects, the planets would likely plunge straight into the sun because of the mutual gravitational forces between them and the sun.

Let's say the sun and Jupiter are the two dots shown in Figure 12. Jupiter and the sun are equivalent to 318 and 333,000 times the mass of the earth respectively. The average distance between the sun and Jupiter is 778.3 million kilometres (km). The mutual gravitational force between the sun and Jupiter is $4.17023 \cdot 10^{23}$N. The computation for the mutual gravitational forces between the sun and Jupiter is shown below:

$$F = G \, M_{sun} \, M_{Jupiter} / r$$
$$= (6.67259 \cdot 10^{-11}) \text{Nm}^2\text{kg}^{-2} (1.991 \cdot 10^{30}) \text{kg} (1.90 \times 10^{27}) \text{kg} / [(7.78 \cdot 10^{11}) \text{m}]^2$$
$$= 4.17023 \cdot 10^{23} \text{N}$$

There is a relatively strong mutual attraction between the sun and Jupiter, enough to send Jupiter plunging into the sun. But this has never happened. Jupiter continues to revolve around the sun. Why?

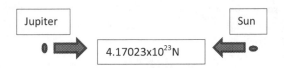

Figure 12: Gravitational force between Jupiter and the sun

The calculation above assumes that the initial condition for the sun and Jupiter is static. But Jupiter is not static. We know that Jupiter makes a complete revolution around the sun in 11.86 earth years. That is a tremendous speed in view of Jupiter's large mass.

We reckon that the initial distance between the sun and Jupiter was much farther than $7.78 \cdot 10^{11}$m. So Jupiter would drift towards the sun and gradually increase its speed. As Jupiter moved closer to the sun, it felt the pull of the sun's spinning-gravitational force grow stronger. It would start to revolve around the sun at the distance of $7.78 \cdot 10^{11}$m as shown in Figure 13.

We cannot assume the sun is simply a dot, because the sun is not static. It revolves on an axis. Likewise, the relative orbits of the earth and moon come into existence because of the spinning-gravitational effects of the earth on the moon. Otherwise the moon would have plunged into the earth before the moon had a chance to establish an orbit.

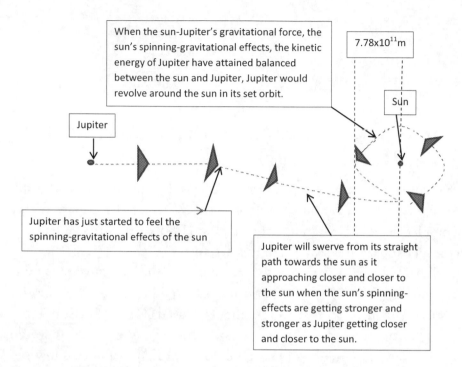

When the sun-Jupiter's gravitational force, the sun's spinning-gravitational effects, the kinetic energy of Jupiter have attained balanced between the sun and Jupiter, Jupiter would revolve around the sun in its set orbit.

$7.78 \times 10^{11}$m

Sun

Jupiter

Jupiter has just started to feel the spinning-gravitational effects of the sun

Jupiter will swerve from its straight path towards the sun as it approaching closer and closer to the sun when the sun's spinning-effects are getting stronger and stronger as Jupiter getting closer and closer to the sun.

Figure 13: The sun's spinning-gravitational effects on Jupiter

The spinning-gravitational effects of the sun can also help us to understand the formation of the solar system. Table 2 shows the inclination to the ecliptic and the inclination to the sun's equator of all planets in the solar system.

A planet's angle of inclination to the sun's equator tells us the possible initial direction of the planet while it was drifting towards the sun, before settling in its current orbit. For instance, the earth's inclination to sun's equator is 7.155° as shown in Table 2. The possible initial location of the earth can be estimated as illustrated in Figure 14.

The inclination to the ecliptic can also help us to figure out which portion of the celestial sphere was the possible initial location of the planet in question. This information is crucial, as it helps determine which planets may have had a shared origin point and thus could possibly be made of similar material.

|  | Name | Inclination to ecliptic(degree) | Inclination to sun's equator(degree) |
|---|---|---|---|
| Terrestrials | Mercury | 7.01 | 3.38 |
|  | Venus | 3.59 | 3.86 |
|  | Earth | 0 | 7.155 |
|  | Mars | 1.88 | 5.65 |
| Gas Giants | Jupiter | 1.31 | 6.09 |
|  | Saturn | 2.49 | 5.51 |
|  | Uranus | 0.77 | 6.48 |
|  | Neptune | 1.77 | 6.43 |

Table 2: Planetary inclination to ecliptic and inclination to sun's equator

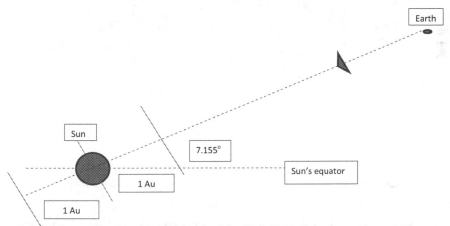

Figure 14: The possible origin of the earth based on info of inclination to the sun's equator

Based on the information in Table 2, we can safely claim that the solar system's planets must have originated from different portions of celestial sphere, even before we classify them into two distinct groups, namely terrestrials and gas giants. The argument is that astronomers use arc second in parallax to determine the distance to a star. A difference of more than 0.1° equals 360 arc seconds, which is considered a huge difference in inclination to the ecliptic. Of course, in our case we are trying to determine the possible original location of a planet, rather than the distance to a star. Still, more than 0.1° in inclination to the ecliptic is considered a significant difference in astronomical terms.

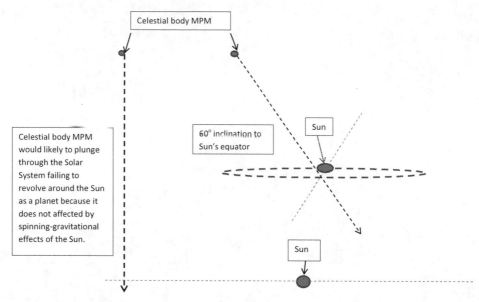

Celestial body MPM

Celestial body MPM would likely to plunge through the Solar System failing to revolve around the Sun as a planet because it does not affected by spinning-gravitational effects of the Sun.

60° inclination to Sun's equator

Sun

Sun

Figure 15: Celestial body with huge inclination to sun's equator will likely to plunge through the solar system

Referring to Table 2, we reckon that any somewhat massive celestial body that approaches the sun within plus or minus 30° in inclination to the sun's equator will likely experience the sun's spinning-gravitational effects such that the celestial body would become a planet in the solar system (as broadly defined above to include dwarf planets and asteroids).

Anything that is less massive, such as comets, whose inclination to the sun's equator is more than ± 30° would enter orbits moulded mainly by pure gravitational force, with minimal influence from the spinning-gravitational effects of the sun. When less massive bodies approach the sun, they may not settle in a permanent orbit and may get too close, causing them to plunge into it.

When we consider an exoplanet already settled in a permanent orbit in a system, we have to employ three factors: the mutual gravitational attraction between the planet and the star, the spinning–gravitational effects of the star, and the velocity of the planet. Normally the inclination of the planet to the star should be within ± 30°.

Let's say there was a massive celestial body named MPM at a 60° inclination to the sun's equator and approaching the sun, as shown in Figure 15. Since its inclination is much greater than ± 30°, it would not effectively experience the sun's spinning-gravitational effects

In this situation, there are three possibilities. The first possibility is that MPM would move too close to the sun, likely triggering a plunge into the sun. The second possibility is that MPM would move too far away from the sun, likely causing it to drift away. The third possibility is that MPM would settle in a temporary orbit. In this last case, the spinning-gravitational effects of the sun would have minimal influence in moulding the orbit of MPM. Mutual gravitational attraction and MPM's velocity would determine the nature of its orbit. A stable orbit would result once in a blue moon. The erratic velocity of MPM would most likely cause it to move closer to the sun or drift farther away.

We cannot assume a star like the sun, a planet like the earth, or even an enormous body like a galaxy to be a point mass. Any celestial body rotating on its own axis has spinning-gravitational effects on other celestial bodies at its vicinity, which enables planetary systems like the solar system to be formed.

By the same token, it is possible for the moon to revolve around the earth or for a spiral galaxy to gyrate around gigantic stars in its central bulge. Of course, the wobble of the sun makes its spinning-gravitational effects even more prominent and effective.

## Prediction of a Massive Planet Nine Is a Bluff

Some astronomers have predicted the existence of a massive Planet Nine in our solar system. It is believed to be as massive as Uranus and to follow an extremely large orbit around the sun. Is this prediction sensible?

It is sensible to expect a massive celestial body to revolve around the sun in a large orbit, because there would be mutual gravitational force between sun and planet despite their distance. The mutual gravitational force between Planet Nine and the sun would nevertheless be very weak when they are farthest apart. The interesting question would be how such a weak mutual gravitational pull could resist the massive momentum of a

Uranus-size celestial body and keep it in its orbit. A so-called Planet Nine would most likely drift away from the sun permanently due to its massive momentum.

## Existence of Gravitational Waves Is a Bluff

Einstein predicted the existence of gravitational waves, which all celestial bodies are believed to rest on. The Laser Interferometer Gravitational-Wave Observatory (LIGO) recently claimed that it has detected the presence of gravitational waves for a second time as the result of a collision between two massive black holes.

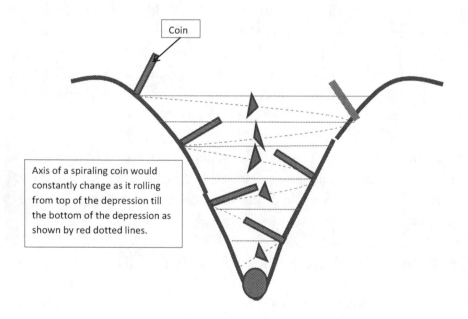

Figure 16: A coin rolling down a depression

A dense star like the sun that is sitting on a gravitational wave would make a depression in that wave. The more massive the star, the deeper it would sink.

We can construct a geometrical depression in laboratory, like the one shown in Figure 16, that is similar to the one made by a star on a gravitational wave. Then we could roll a coin from the top of the depression so that it spirals downward to the bottom of the depression.

The axis of the coin is constantly changing as it rolls down the depression, as shown in Figure 16. So we can expect the axis of a planet that is revolving around a star to constantly change in a similar way.

The earth's rotational axis slants at 23.44°. The earth maintains this slant as it revolves around the sun, as shown in Figure 17. The slanting rotational axis of the earth gives rise to seasonal changes.

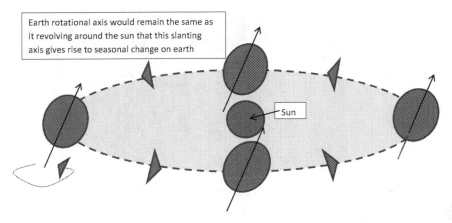

Figure 17: Earth maintain its rotational axis while revolving around the sun

If gravitational waves existed, then the rotational axis of the earth would have to change constantly, as Figure 16 illustrates. That is manifestly not the case.

The universe is still expanding, and all celestial bodies are still mutually attracted. In consequence, they constantly adjust as they all shift. Therefore we should detect constant tremors on earth due to these adjustments. This doesn't necessarily imply the existence of the gravitational waves.

## Accumulative Masses Give Rise to Planetary Rotational Axis

The planets in the solar system do not have the same axial tilt. Nor do they all share the same rotational direction. Why?

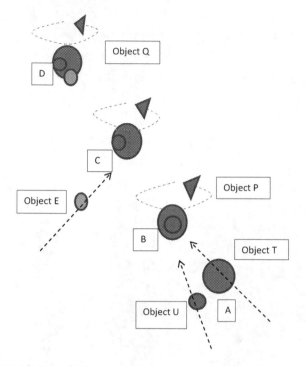

Figure 18: Accumulative masses give rise to planetary rotational axis

All stars, including the sun, and all planets in our solar system were built up piece by piece out of smaller building blocks shortly after the Big Bang explosion. Hot objects colliding were likely to coalesce.

When bigger, massive Object T and smaller, lighter Object U collided, they coalesced into Object P. Object P started to rotate in a particular direction, as shown in Figure 18, determined by the conservation of the momentums of Objects T and U. Object P became the seed of a much bigger boulder. Subsequent accumulation turned it into a planet or, under the right circumstances, a much larger celestial body, like a star.

The crucial information of the initial momentum of Object T and Object U are important in helping us to deduce the rotational direction after they have coalesced into one, based on the principle of the conservation of momentum.

In Figure 18 again, another Object E drifted towards Object P and collided with it. The two coalesced into Object Q. How Object Q tilts depends upon where the Object E strikes Object P. How Object Q continues to rotate also depends upon how Object E strikes Object P.

Of course, the construction of a planet from smaller building blocks is a rather random process. Each planet in the solar system has a different mass, size, rotational speed, rotational direction, and tilt.

| Planets | Axial tilts of the planets |
|---------|----------------------------|
| Mercury | 0° |
| Venus | 178° |
| Earth | 23.44° |
| Mars | 25.20° |
| Jupiter | 3.12° |
| Saturn | 26.73° |
| Uranus | 97.86° |
| Neptune | 29.56° |

Table 3: Planetary axial tilts in the solar system

Once a protoplanet or star has gained enough mass, then its spinning-gravitational effect on its rotational axis will be strong enough to assist in its further construction. It will conserve momentum while continuing to attract adjacent objects. Attracted objects will revolve around it in parallel to its rotational direction before plunging into it.

Only Venus rotates in a counterclockwise direction; the rest of the planets have a clockwise rotational direction. Each planet has a different axial tilt, which strongly suggests that a random process was dominant in the construction of our solar system's planets.

## Planets and Radioactive Substances

Generally, planets are made out of a mixture of five types of materials: very light radioactive substances (lithium to zirconium), light radioactive substances (niobium to radium), dense radioactive substances (actinium to atomic number 129), very dense radioactive substances (atomic number 130 and above) and non-radioactive substances. Some planets may have more light radioactive substances than dense and very dense radioactive substances. Others may have a relative abundance of very light radioactive substances.

To make things more complicated, the daughter elements of some dense radioactive substances are very light or light radioactive substances. Likewise, the daughter elements of some very dense radioactive substances are dense, light, or very light radioactive substances.

Light and very light radioactive substances may transform and release moderate amounts of heat. The nuclei of dense and very dense radioactive substances can be split, but the nuclei of light and very light radioactive substances can only undergo minor restructuring without splitting their nuclear structure.

There are many elements that can be categorized as light radioactive substances. There are also a lot of elements that can be classified as dense and very dense radioactive substances. A particular radioactive element can also have many different radio isotopes, not all of the same density.

In our solar system, every planet consists of a mixture of very light and light radioactive substances. Some of them may also have quite a lot of dense radioactive substances. Astronomers agree that each planet in the solar system has a core and a mantle. This implies that each planet was once a lump of hot materials. Evidence that this is true comes from the fact that all of them are spheroid in shape. Gradual cooling would allow a mass to form a spheroid.

Rotational speed and gravitational pull also gradually shape the planet. For instance, the earth's rotational speed is considered slow at twenty-four hours per rotation, and its gravitational pull is considered even weaker at 9.8 kgm/s$^2$. Given its physical size, its rotational speed is huge compared to its gravitational pull. The disparity between these forces causes the earth to bulge at its equator.

A lot of astronomers believe that Jupiter has no clear-cut surface. Storm clouds shroud Jupiter, obscuring our view and giving rise to this uncertainty as to whether there is a true surface or not. However, just because we can't see it directly doesn't mean that there is no surface.

Jupiter's rotational speed is considered very fast at 9.93 hours per revolution, especially in view of its enormous size. Yet Jupiter does not bulge at its equator, mainly because of its enormous gravitational pull. The presence of such a gravitational pull would surely allow the formation of a surface. Therefore we strongly believe that Jupiter has a surface.

After the impact of the Shoemaker–Levy 9 comet on Jupiter, a splash of debris was flung through the storm clouds and settled back again. If Jupiter didn't have a surface, then Shoemaker–Levy 9 would have punched straight towards the core of Jupiter without causing a splash.

Jupiter is far from the earth and even farther from the sun, yet its surface temperature is relatively high due to heat from its core, which is masked by its shroud of clouds. Its temperature must be high; otherwise the ammonia gas in its atmosphere could not be produced by the reaction of hydrogen and nitrogen gases under the presence of tremendous Jupiter's atmospheric pressure. Therefore, we believe that Jupiter's atmospheric temperature is much hotter than the earth's but much lower than the sun's.

NASA has sent probes to all the planets in the solar system. We hope the temperatures recorded by those probes are correct. The surface temperature of a planet is influenced by its exposure to direct sunlight, which in turn depends on how far the planet is from the sun. The relative absence of sunlight dramatically lowers temperature. Another factor is whether the planet itself is emitting heat. The presence of a cloud system also has an effect, reflecting heat back to the surface.

Based on the atmospheric composition of a planet, we can predict what materials used to be in its core. This is because, after a sequence of decay, emitted particles, daughter elements, and their compounds end up in the atmosphere.

For instance, hydrogen gas in the atmosphere is derived from the radioactive decay of dense radioactive substances in the core, since dense radioactive substances are most likely to undergo either proton or neutron emission. Sometime very dense radioactive substances emit hydrogen gas during their decay, but this is relatively rare.

The presence of helium in the atmosphere shows that the core of a planet used to be filled with very dense radioactive substances, since they tend to undergo alpha emission during decay. Sometime dense radioactive substances also undergo alpha emission, but again this is relatively rare.

Based on current atmospheric composition, surface temperature, and volcanic activity, scientists can also predict the types of radioactive substances that are still embedded in the core of the planet.

# Mercury

Space probe Messenger, on 29 September 2009, discovered the crater of a volcano on Mercury. Mercury appears to have experienced a high level of volcanic activity, strongly suggesting the presence of water in its past. Pressurized steam, or water vapor, is what gives rise to volcanic eruptions.

Table 4 shows the atmospheric composition of Mercury as measured by Messenger space probe.

There is no longer volcanic activity on Mercury. The core of Mercury must be rather cool nowadays. Therefore it is safe to claim that Mercury's core is primarily filled with very light radioactive substances, which do not emit much heat during their decay.

There is an obvious lack of nitrogen in Mercury's atmosphere. Likely there are no basic life forms on Mercury, because nitrogen is the crucial building block of proteins.

There are small concentrations of both hydrogen and helium gases present in Mercury's atmosphere, implying that it used to have both dense and very dense radioactive substances in its core.

| Element | Column Density ($cm^{-2}$) |
|---|---|
| Hydrogen | $\approx 3 \times 10^9$ |
| Molecular Hydrogen | $< 3 \times 10^{15}$ |
| Helium | $< 3 \times 10^{11}$ |
| Atomic oxygen | $< 9 \times 10^{14}$ |
| Sodium | $\approx 2 \times 10^{11}$ |
| Potassium | $\approx 2 \times 10^9$ |
| Calcium | $\approx 1.1 \times 10^8$ |
| Magnesium | $\approx 4 \times 10^{10}$ |
| Argon | $\approx 1.3 \times 10^9$ |
| Water | $< 1 \times 10^{12}$ |
| Neon, silicone, sulfur, argon, iron, carbon dioxide, etc. also been detected | |

Table 4: Mercury's atmospheric composition

The presence of atomic oxygen in Mercury's atmosphere shows that there are not a lot metallic elements present, since the atomic oxygen has escaped the core without reacting to form metallic oxides. Likewise, there is a small amount of carbon in the atmosphere. If metallic oxides were present, the carbon would have reacted with the oxygen atoms in the metallic oxide compounds to form carbon dioxide.

The presence of metallic atoms, like sodium, potassium, calcium, and magnesium, in Mercury's atmosphere could be due to extremely high temperatures as a result of the planet's proximity to the sun.

Apart from direct, intense sunlight on its daylight side, the surface temperature of Mercury is actually rather cold, about -170° C on its night side. This implies that the core of Mercury must still be filled with very light radioactive substances that do not emit much heat.

## Venus

Scientists have detected indications of relatively recent lava flows on the surface of Venus. Venus is one of the few worlds in our solar system that has been volcanically active in the last three million years. Volcanic activity is a clear sign that water was present on Venus.

Table 5 shows the atmospheric composition of Venus as measured by NASA's probe.

| Compounds | Measurements |
|---|---|
| Carbon dioxide | 96.5% |
| Nitrogen molecule | 3.5% |
| Sulfur dioxide | 150ppm |
| Argon | 70ppm |
| Water vapor | 20ppm |
| Carbon monoxide | 17ppm |
| Helium | 12ppm |
| Neon | 7ppm |
| Hydrogen chloride | 0.1-0.6ppm |
| Hydrogen fluoride | 0.001-0.005ppm |

Table 5: Venus's atmospheric composition

There are trace amounts of helium and hydrogen compounds in Venus's atmosphere, which attests that the core of Venus used to be filled with corresponding amounts of very dense radioactive substances and dense radioactive substances.

Venus's surface temperature does not vary much, even at night, due to its high carbon dioxide concentration that traps heat very well. The average temperature on Venus is 470° C, which indirectly shows that the diffusion of gases is very fast between the atmosphere that constantly faces the sun and the atmosphere that always faces the night. We reckon that the core of Venus is still filled with much lighter radioactive substances, since there is no volcanic activity on Venus any longer.

The presence of a small amount of carbon monoxide and an abundance of carbon dioxide proves that a lot of carbon and oxygen were released as daughter elements from the decay of dense radioactive substances. In addition, the presence of argon and neon gases, the daughter elements of the decay of a particular type of dense radioactive substance. The very light radioactive substances in Venus's core are likely still undergoing beta emissions and dissipation of low intensity of heat as they decay.

## Earth

The earth's atmospheric composition is shown in Table 6. The vast ocean on earth tells us that there is an abundance of hydrogen and oxygen, the daughter elements of the decay of dense radioactive substances in the core. A trace amount of helium in the upper atmosphere shows that there also used to be a small amount of very dense radioactive substances. Carbon and nitrogen in the earth's atmosphere are also the result of the decay of mostly dense radioactive substances. Earth's mantle and core are still rather hot, which strongly suggests that there are still plenty of dense radioactive substances in the core.

The daughter elements of decaying dense radioactive substances are mostly heavy metallic elements and some non-metallic and light metallic compounds. Dense radioactive substances reduce to light radioactive substances, very light radioactive substances, or both when they undergo decay.

The daughter elements of decaying light radioactive substances are heavy metallic elements and their compounds. The daughter elements of decaying much lighter radioactive substances are some light metallic and non-metallic elements and their compounds.

Generally, oxygen atoms released as the daughter element of decaying dense radioactive substances subsequently react with metallic elements to form metallic oxides. Carbon and hydrogen atoms reduce oxygen anions in metallic oxide compounds to form carbon dioxide and water respectively. This is how water and carbon dioxide have been produced on earth. This understanding of how water comes into existence makes us believe that the existence of water in the universe is rather common.

Due to the existence of plants, carbon dioxide has been gradually converted to oxygen through photosynthesis. Therefore, the concentration of oxygen on earth is relatively high. The concentration of nitrogen gas is also rather high, reducing the volatility of the oxygen. Otherwise there would be a high rate of oxidation in living cells.

Nitrogen atoms, as a daughter element of radioactive decay in the core, do not react with anything they come into contact with due to their inert characteristics. They eventually surface in the earth's atmosphere as nitrogen gas.

| Gas name | In % |
|---|---|
| Nitrogen | 78.084 |
| Oxygen | 20.946 |
| Argon | 0.9340 |
| Carbon dioxide | 0.0397 |
| Neon | 0.001818 |
| Helium | 0.000524 |
| Methane | 0.000179 |
| Water vapor | 0.001-5 |

Table 6: Earth's atmospheric composition

The presence of uranium in the earth's crust signals that the core must still be filled with a sizeable amount of dense radioactive substances. The continuous release of radioactive carbon-14 in the form of carbon dioxide could also be the result of the radioactive decay of those dense radioactive substances. Volcanic activity is still high on earth and could likewise be energized by the presence of dense radioactive substances in the core.

A big portion of earth's mantle and core are still filled with very light and light radioactive substances that are dissipating low-intensity heat as they decay. We reckon 20 per cent of the earth's material consists of non-radioactive substances.

## Mars

NASA's scientists predict that there is a considerable amount of water ice present in Mars's subsurface. Traces of water, methane, and sulphur dioxide have been detected in the atmosphere. Ozone concentration is said to be 300 times lower than that of the earth. The atmospheric composition of Mars is shown in Table 7. Surprisingly, there is abundance of carbon dioxide.

The absence of helium in Mars's atmosphere tells us that there weren't any very dense radioactive substances present in the core. Trace amounts of hydrogen compounds, like water ice and vapor, strongly suggest that a small amount of dense radioactive substances used to be present in the core.

| Name of Compounds | In % |
|-------------------|--------|
| Carbon dioxide | 95.97 |
| Argon | 1.93 |
| Nitrogen | 1.89 |
| Oxygen | 0.146 |
| Carbon monoxide | 0.0557 |

Table 7: Martian atmospheric composition

The absence of plants means there is no mechanism to convert carbon dioxide to oxygen. Even if there were plants on Mars, these would fail to support other life forms because of the low nitrogen concentration. Without nitrogen to dilute the oxygen, the atmosphere would be very volatile. If life forms exist on Mars, they are unexpected ones that can harness carbon dioxide.

The atmospheric composition of Mars is similar to that of Venus. This leads us to believe that Mars's core and Venus's core must have been filled with similar types of dense radioactive substances. Carbon and oxygen have been produced as the daughter elements of the decay of dense radioactive substances. The core of Mars is likely still filled with mainly very light radioactive substances, since there is a complete absence of volcanic activity.

## Asteroids

The space between orbit of Mars and Jupiter is filled with asteroids. It is believed that there were two planets of almost equal mass, set in almost equal orbits, that collided. The collision reduced the two planets into fragments that we call asteroids, as shown in Figure 19.

We don't expect such planets to have a straight collision while they are in their orbits. Let's say that initially, the smaller planet PlntAstrd1 is getting rather close to the larger planet PlntAstrd2 while revolving around the sun. Once PlntAstrd1 enters the gravitational field of PlntAstrd2, PlntAstrd1 will divert from its own orbit and start to circulate around PlntAstrd2 under its spinning-gravitational effect. As the time goes by, PlntAstrd1 gets closer to PlntAstrd2. Eventually, PlntAstrd1 will plunge into PlntAstrd2. Assuming both planets solidified before colliding, the resulting fragments would be irregular rather than spheroid.

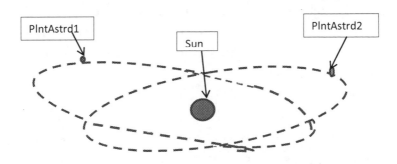

Figure 19: The collision between two huge asteroids leading to the formation of asteroid belt

Since asteroids are cool and not volatile, we reckon that both PlntAstrd1 and PlntAstrd2 consisted of mostly much lighter radioactive substances and some light radioactive substances that derived from dense radioactive substances. Asteroids may be enriched with mostly non-metallic compounds and light metallic compounds, and some heavy metallic compounds. We can't make assumptions about asteroid composition based on the composition of meteorites, because almost all active light and heavy metallic elements that may have existed in meteorites would have been oxidized during their entry into the earth's atmsphere.

## Jupiter

Jupiter's atmospheric composition is enriched with hydrogen, helium, and trace compounds, as shown in Table 8. We reckon that Jupiter's core used to be filled with a lot of very dense radioactive substances and some dense radioactive substances.

Surprisingly, Jupiter is not any hotter than the sun. This implies that the core of Jupiter must now be filled mostly with light radioactive substances and some dense radioactive substances. Light radioactive substances release a low intensity of heat during their decay. Dense radioactive substances release intense heat.

Unlike the sun, the core of which is almost 50 per cent dense radioactive substances, Jupiter has about 20 per cent dense radioactive substances. Thus, Jupiter releases much less intense heat than the sun.

If Jupiter's core were still filled with dense radioactive substances, we would expect it to glow bright and hot like the sun. It would, in effect, be an additional dwarf star in the solar system.

Due to Jupiter's extremely high atmospheric pressure and rather high temperature, chemically inert nitrogen gas reacts with hydrogen gas to form ammonia gas. Jupiter's aerosol clouds are laced with ammonia ice, water ice, and ammonia hydrosulphide. We estimate that Jupiter's surface temperature could be as high as 1,500° C.

Composition by volume (uncertainty in parentheses)

| Name | Measurements |
|---|---|
| Hydrogen | 89.89%(2%) |
| Helium | 10.29%(2%) |
| Methane | 3000ppm(1000ppm) |
| Ammonia | 260ppm(40ppm) |
| Hydrogen Deuteride | 28ppm(10ppm) |
| Ethane | 5.8ppm(1.5ppm) |
| Water | 4ppm(varied with pressure) |
| Aerosols (ammonia ice, water ice, ammonia hydrosulfide) | |

Table 8: Jupiter's atmospheric composition

The Great Red Spot on Jupiter is a stormy region similar to a hurricane on earth. A hurricane on earth can only last for a few weeks at most, and quickly subsides once it makes landfall. This is because land doesn't support convection in the way relatively warm sea water does.

The conditions that give rise to the Great Red Spot, on the other hand, could be sustained for centuries. This make us believe that there must be something that keeps the storm constantly brewing. Maybe there is a gigantic active volcano underneath the storm clouds. The continuous spewing of sulphur from an active volcano would also account for the colouring of the Great Red Spot.

As a matter of fact, there must be plenty of volcanoes on Jupiter and the biggest volcano in Jupiter must be the one underneath the Great Red Spot. Therefore we can expect a lot of active volcanic activities on Jupiter. Super earthquakes are undoubtedly a common occurrences on Jupiter.

The surface temperature of Jupiter must be rather high, hot enough to evaporate all the surface water despite the high atmospheric pressure. Jupiter is smothered in very dense clouds laced with ammonia ice and ammonia hydrosulphide. Since the upper atmosphere of Jupiter is relatively cold, condensation in those clouds takes place. Therefore, rain is fairly common on Jupiter. Jovian thunderstorms are on a much greater scale than the ones on earth. Dense clouds, strong winds, and high temperatures are a perfect environment to constantly brew super thunderstorms.

Very dense clouds conserve Jupiter's high surface temperature by efficiently conserving the low-intensity heat dissipating from those same clouds. We can see Jupiter glowing in the viewfinder of an infrared telescope.

It is wrong to classify Jupiter as a gas giant, because it is not simply a lump of hydrogen gas. The Shoemaker–Levy 9 comet impact proved that Jupiter has a surface. Jupiter has a surface partly because of its strong gravitational pull. We strongly believe that the core of Jupiter is still filled with mostly very light and light radioactive substances, along with a sizeable amount of dense radioactive substances, in view of its rather high temperature and strong magnetic field.

## Saturn

Saturn's atmosphere consists of 96.3 per cent molecular hydrogen and 3.25 per cent helium by volume. This implies that the core of Saturn used to be filled with dense radioactive substances, along with a small amount of very dense radioactive substances. Trace amounts of ammonia, acetylene, ethane, propane, phosphine, and methane have been detected by NASA's probes.

Like Jupiter, Saturn's atmospheric pressure is high. It also has a rather high atmospheric temperatures. Thus, its atmospheric mixture of hydrogen and nitrogen churns out ammonia. Its upper clouds are laced with ammonia crystals, and its lower clouds are laced with ammonium hydrosulphide.

Like Jupiter, Saturn is smothered in very dense clouds. Its surface temperature should be hot enough to evaporate any water, despite its high atmospheric pressure. We reckon that there are also a lot of volcanoes on Saturn that are constantly brewing stormy conditions, especially at the Great White Spot. The dense cloud system conserves the heat energy dissipating from those clouds.

Saturn is as hot as Jupiter. But both of them are much colder than the sun. Therefore we reckon that Saturn's core must still be filled with sizeable amount of dense radioactive substances that release a substantial number of dynamic photons during their radioactive decay. The core of Saturn may contain lesser quantities of dense radioactive substances (about 15 per cent) than the core of Jupiter, given that Saturn has a much weaker magnetic field. A big portion of Saturn's core is composed of very light and light radioactive substances, derived from decaying dense radioactive substances. Like Jupiter, Saturn should not be classified as a gas giant because it is composed mostly of radioactive materials, not gas.

## Uranus

Uranus's atmosphere contains roughly three quarters hydrogen and a quarter helium by mass. This implies that Uranus's core used to be filled with dense radioactive substances. In addition, some very dense radioactive substances were present.

NASA's probes have also detected a small amount of methane and trace amounts of water, ammonia, and other substances. The highest clouds are very bright and laced with frozen methane. Farther down, there are perhaps clouds composed of frozen water and ammonium hydrosulphide.

Like Jupiter, Uranus's atmospheric pressure and temperature are rather high and could produce ammonia from daughter elements hydrogen and nitrogen. High temperatures also constantly brew stormy conditions on Uranus. These are less severe than the storms on Jupiter, Saturn, and Neptune. Therefore, we expect the surface temperature of Uranus to be colder.

Uranus, like Jupiter and Saturn, is smothered in clouds, but their density is not as great. The clouds on Uranus appear relatively plain and featureless. There may be fewer volcanoes on Uranus than on Jupiter, and apparently those volcanoes are less active.

Uranus appears to be much cooler than Jupiter but much hotter than the earth. Its core lkely contains about 10 percent dense radioactive substances, in addition to a big portion of very light and light radioactive substance. Like Jupiter, Uranus is not simply a gas giant.

## Neptune

Neptune's atmosphere is 80 per cent hydrogen and 19 per cent helium as recorded by NASA's space probes. This implies that the core of Neptune used to contain an abundance of dense radioactive substances and a sizeable amount of very dense radioactive substances. There is trace of methane. The upper troposphere is laced with ammonia and hydrogen sulphide. The lower troposphere is laced with ammonia, ammonium sulphide, and water.

Like Jupiter, Neptune's atmospheric temperature and pressure are rather high and could produce ammonia. Neptune is smothered in dense clouds. The high surface temperature gives rise to stormy conditions. The Great Dark Spot on Neptune could be similar to the Great Red Spot on Jupiter, supported by a gigantic volcano underneath it.

Neptune's lower stratosphere shows the presence of ethane, ethyne, carbon monoxide, and hydrogen cyanide. Carbon dioxide and water in thermosphere have been detected by NASA's space probes.

Like Jupiter, we reckon the core of Neptune contains about 13 per cent dense radioactive substances. A big portion of the core consists of very light and light radioactive substances that are likely to dissipate low-intensity heat. Neptune is much cooler than Saturn, but much hotter than Uranus. As with Jupiter and the other giant planets, Neptune shouldn't be categorized as gas giant.

## Kuiper Belt

Understanding the composition of the Kuiper belt may not necessarily help us to understand the formation of solar system. As a matter of fact, the Kuiper belt is made up of very diverse materials. Every dwarf planet that forms the Kuiper belt may have originated from a different portion of the celestial sphere, since each has a different inclination to the ecliptic.

Most dwarf planets in the Kuiper belt are smaller than Pluto, but some may be slightly bigger. Since they are far from the sun, the mutual gravitational pull between them and the sun is rather weak. It is their small mass that keeps them circling the sun. If they were more massive, the mutual gravitational pull would still be too weak to resist their momentum, and they would drift out of their orbits.

# Comets

Generally, a comet's orbit is somewhat eccentric, as shown in Figure 20. When a comet gets close to the sun, the sun's intense heat vaporizes the part of the comet that is directly facing the sun. This dust forms a characteristic long tail.

Astronomers believe that comets contain water. Specifically, they think that vaporizing ice in the comet allows the formation of the tail. This leads to the common description of a comet as a "dirty snowball".

Astronomers may have underestimated the power of very intense sunlight, which could vaporize anything if given sufficient time. The tail may not necessarily be due to the vaporizing of ice.

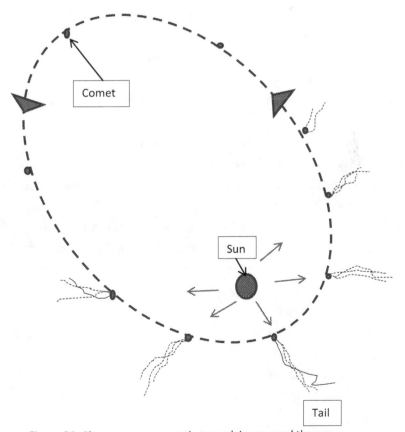

Figure 20: Changes to a comet that revolving around the sun

Some astronomers believe that comets falling on earth originally brought water to our planet. They even believe that there was plenty of water at the edge of the solar system, and comets were tasked to trickle water everywhere inside the solar system. This view seems to presume that comets are like super sponges, effectively soaking up even frozen water.

## How the Moon Came into Existence

Both the moon and the earth are spheroid in shape. There are no signs of severe chips in them. This implies they have experienced gradual cooling since their formations, and that no big celestial bodies collided with them after they solidified.

If a Mars-like object had collided with the earth while the earth was still a hot lump, most likely the two celestial objects would have coalesced after sending out some splashes.

Though terran and lunar rocks share some similarities in term of their mineral content, these similarities don't necessarily imply that the moon was a chip off the earth. There are similarities and differences in every nook and cranny of the universe.

After the earth settled into its orbit around the sun, the moon started to drift towards the sun from afar. Like a comet, the moon revolved around the sun in an eccentric orbit, but that orbit was somehow intercepted by the earth. Subsequently, the moon settled into an orbit that revolved around the earth, under the influence of the earth's spinning-gravitational effects.

NASA claims that the moon is drifting four inches farther away from the earth every year. If this is true, it implies that the earth is revolving too close to the sun. The moon must have excess kinetic energy. If the earth's orbit were slightly farther away from the sun, the moon might settle into a more stable orbit. As things are, if NASA's claim is correct, the moon will eventually drift away from the earth. The moon might ultimately become like a comet again, plunging closer and closer to the sun while revolving around it. Whether the moon will plunge into the sun is yet to be known.

## Multiverse Is a Bluff

An atom can resist pressure imposed on it mainly because the existence of the mosaic of magnetic fields of the unique nucleus structure, which probe the stationary electrons and prevent them from collapsing into the nucleus.

Those stationary electrons are revolving around the nucleus at very high velocities. Without the magnetic fields, any force acting on the stationary electron would cause it to plunge straight into the nucleus.

All the celestial bodies in the universe are drifting out from the centre. There is no mechanism probing them and keeping them from collapsing, other than their kinetic energy.

If the multiverse ever existed, with several universes resting on top of one another, this would surely cause the implosion of all the universes. The universes would have no mechanism to resist collapse and would be flimsy against external forces.

Luckily, this is not what has happened. Therefore it is safe to say that the multiverse is a bluff. There is only one universe, and that is our universe.

## The Claim that Nothing Can Travel Faster than Light Is a Fallacy

NASA scientists claim that the Hubble telescope has visualised the nascent state of the universe. In other words, we can really look at the universe when it was in its formative stage. If this is true, it implies that all celestial bodies in the universe were drifting much faster than the speed of light during the Big Bang. Subsequently, those celestial bodies slowed down but were still travelling faster than the speed of light.

The universe is still expanding, but it is expanding at a much slower pace as compared to its past record. Nowadays the expansion is slow enough for the light of the past to catch up with us on earth. That is what enables us to see the formative phase of the universe. Our assumption is that the speed of light is a constant, $3 \cdot 10^8$ m/s.

If celestial bodies travelled much faster than the speed of light during the Big Bang and shortly after, then it is a fallacy that nothing can travel faster than the speed of light.

## The Existence of Cosmic Background Radiation

The universe will continue to expand until all the kinetic energy of its celestial bodies has completely transformed to universal gravitational potential energy. At that point, the universe will have expanded to its maximum size. The celestial bodies will then stop momentarily before imploding back to the centre of the universe. As these bodies implode from the far edges of the universe, universal gravitational potential energy will convert back to kinetic energy at an accelerating rate. When celestial bodies near the centre of the universe, they reach tremendous speed. As they collide, they will be flung violently out of the centre to start a new Big Bang cycle.

In short, the universe will always be there, and the cycle of expansion and implosion will repeat forever.

In every expansion, some dynamic photons travel far beyond the edge of the universe. Like other celestial bodies, these dynamic photons will cease to move when all their kinetic energy has transformed to universal gravitational potential energy. Then they will start to implode.

The exchange of photons is a constant process. We believe that celestial bodies will continue to exchange photons as they implode, and that some dynamic photons from the imploding celestial bodies will head far, far beyond the edge of the universe. At some point, these too will implode. The perennial imploding photons from beyond the edge of the universe are what we have named *ghost photons*. The existence of ghost photons would give rise to cosmic background radiation.

## The Magnetic Field of a Celestial Body

Table 9 shows that Mars has the weakest magnetic field among the planets, and Venus has the second weakest. Mercury's magnetic field is much stronger. The earth's magnetic field is ranked fourth weakest after Mercury.

Among the giant planets, Jupiter has the strongest magnetic field, and Uranus has the second strongest. Neptune has the weakest magnetic field, and Saturn is the second weakest. The earth's magnetic field is weaker than Jupiter's but slightly stronger than that of Uranus.

Since magnetic field strengths are different among the Jovian planets, we can expect the occurrence of auroras with different intensities on them. Jupiter has the most intense auroras near its north magnetic pole, and Saturn has the second most intense. Auroras above Neptune's north magnetic pole are less intense than the ones on Saturn. Uranus's auroras have the weakest intensity of all. This hints that Uranus may actually have the weakest magnetic field strength among the Jovian planets, contrary to the data in Table 9.

Uranus appears to be featureless, which shows that Uranus is not stormier than Neptune and Saturn. Saturn has the Great White Spot and Neptune has the Great Dark Spot, suggesting that Saturn is stormier than Neptune. Based on storm conditions, we expect Saturn to have the highest surface temperature and Uranus the lowest.

|  | Magnetic moment (Earth = 1) | Field at equator (gauss) |
|---|---|---|
| Mercury | 0.0007 | 0.003 |
| Venus | <0.0004 | <0.0003 |
| Earth | 1 | 0.305 |
| Mars | <0.000025 | <0.00005 |
| Jupiter | 20000 | 4.2 |
| Saturn | 600 | 0.2 |
| Uranus | 50 | 0.23 |
| Neptune | 25 | 0.14 |

Table 9 – Planetary magnetic field

The intensity of a planet's aurora depends on two factors, namely magnetic field strength and the supply of dynamic photons. A stronger magnetic field excites atmospheric molecules, causing them to undergo a more intense exchange of photons with their surroundings. That exchange is facilitated if there is an abundant supply of dynamic photons.

We normally witness more intense auroras on the earth when there is also a solar flare. Jovian planets are farther away from the sun, and any flares that reach them are relatively weak. Jovian planets got their supplies of dynamic photons from themselves, specifically from the heat radiating from their cores.

We deduce that Uranus has the weakest magnetic field among the giant planets, since it has the weakest auroras.

We reckon that the cores of Jupiter, Saturn, Neptune, and Uranus are filled mostly with dense radioactive substances. Respectively, there are also about 20 per cent, 15 per cent, 13 per cent, and 10 per cent respectively of dense radioactive substances in their cores. Therefore, their surface temperatures are rather high.

The earth's core contains less than 5 per cent of dense radioactive substances. Its surface temperature is cold, but its mantle and core are hot. The cores of Mercury, Venus, and Mars contain mainly much lighter radioactive substances, and there is no longer volcanic activity on those planets.

Since Mercury's magnetic field is stronger than Venus's, and Mars has the weakest magnetic field, the core of Mercury must be filled with mainly ferromagnetic, very light radioactive substances. Venus's core may be filled with weaker ferromagnetic, very light radioactive substances. Mars's core is likely filled with diamagnetic, very light radioactive substances. We strongly believe the materials in the core of a planet play an important role in determining magnetic field strength.

All Jovian planets have higher surface temperatures than the earth's, so the earth's magnetic field should be slightly weaker than those of the Jovian planets, contrary to the data in Table 9.

A free electron is enveloped with a much stronger magnetic field because it is saturated with more stationary photons. We want to highlight the important role played by stationary photons. A stockpile of stationary photons greatly boosts magnetic field strength. We can draw a parallel between magnetic field strength and the type of radioactive substances in the core of a celestial body. Generally, the nuclei of dense radioactive substances are packed with more stationary photons than the nuclei of much lighter radioactive substances. The amount of dense radioactive substances in the core of a planet determines the magnetic field strength enveloping the planet. Jupiter, with about 20 per cent of dense radioactive substances in its core, is enveloped in a very strong magnetic field.

In conclusion, we have reservations about the accuracy of the magnetic field readings of the planets in solar system, as measured by NASA's space probes. But we agree that there are difficulties in measuring magnetic fields.

## Van Allen Belts

The Van Allen belts are typically categorized as *inner* and *outer*. The inner Van Allen belt is located 1,000 km to 6,000 km above the earth's equator. The outer Van Allen belt is situated between 13,000 km and 60,000 km above the equator. According to NASA, its Van Allen probes have detected a third Van Allen belt, characterized as "temporary" or "transient".

Van Allen believed that the special zones named after him are enveloping magnetic fields that trap charged particles. The inner Van Allen belt traps protons, and the outer Van Allen belt traps protons and electrons. These regions are shaped like doughnuts.

We believe the Van Allen belt on the side of the earth that faces night is smaller and located much closer to the earth, since the sun's activities are weakest. The sizes and locations of the belts change according to the sun's activities. During solar flares, the Van Allen belts shift much farther out from the earth.

NASA's Relativistic Electron-Proton Telescope (REPT), designed to detect protons, has a front lead plate with 200s nm thickness. Underneath the lead plate is a high-performance silicone solid-state detector. To detect electrons, the telescope has a glass filter above the solid-state detector. Readings taken by the proton detector have much higher magnitudes.

The inner Van Allen belt is nested in the exosphere, which is 700 km to 10,000 km above the equator. The exosphere is filled with hydrogen gas, helium atoms, ozone, and nitrogen gas. There is no clear boundary among these gases. Any gas atom or molecule that has gained sufficient stationary photons will drift higher in the Van Allen belts. Gas atoms and molecules move randomly in the Van Allen belts.

Gases in the exosphere, especially between 1,000 km and 6,000 km, stockpile stationary photons very effectively, so they hover in that region in a sustainable manner. Their saturations in the inner Van Allen belt are rather high.

We believe that this is connected with the localized universal gravitational potential energy of atoms. At certain altitudes, the unique nucleus structure flexes in a certain orientation and configuration that may determine how well the atom stockpiles stationary photons. Of course, flexure of the unique nucleus structure in certain manner will re-orient its orbitals.

Gas atoms and molecules that hover in the outer Van Allen belt not only stockpile stationary photons in their nuclei, but their stationary electrons are also saturated with stationary photons. Denser stationary electrons drift farther away from their nuclei, making them easier to ionize. On the other hand, the gas atoms and molecules in the inner Van Allen belt do not amass stationary photons on their stationary electrons. This is partly because of flexure of their unique nucleus structure. The flexure has a certain orientation and configuration due to the atoms' specific, localized, universal gravitational potential energy, prohibiting them from being ionized easily.

Generally, there is higher saturation of gas atoms and molecules in the inner Van Allen belt than in the outer Van Allen belt.

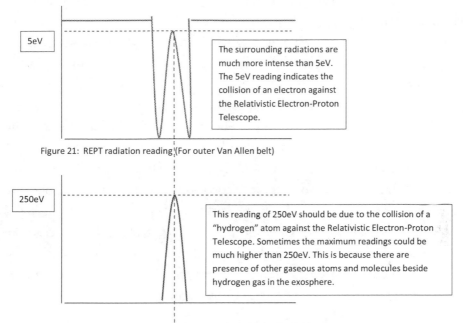

Figure 21: REPT radiation reading (For outer Van Allen belt)

Figure 22: REPT radiation reading (For inner and outer Van Allen belt)

Basically, when the gas atoms and molecules collide against the REPT, they unload dynamic photons to the surroundings. These dynamic photons saturate the high-performance silicone solid-state detector, adhering to the stationary electrons of the silicone molecules and transforming to free electrons. Then they are channelled to the detector to register as energy readings.

Figures 21 and 22 show typical radiation measurements in the outer Van Allen belt. Figure 21 shows that the surrounding radiation is much stronger than 5eV. The 5eV reading is due to the collision of much lighter particles, like the electron, against the REPT. Figure 22 shows much stronger radiation in the outer Van Allen belt. This reading coincides with the occurrence of the 5eV electron. The reason for this reading is the collision of much larger particles against the REPT. Sometimes, this reading is greater than 250eV. This is because of the presence of other gas atoms and molecules beside hydrogen, like helium, ozone, and nitrogen.

The gas atoms in the outer Van Allen belt have nuclei and stationary electrons that are saturated with stationary photons. So when they slam against the REPT, they easily dislodge their stationary electrons. Therefore the reading of 5eV electrons coincides with much higher readings of 250eV for larger particles.

Figure 22 shows typical readings in the inner Van Allen belt, where only readings of 250eV or higher have been recorded. The gas atoms in the inner belt are not amassing stationary photons on their stationary electrons. Therefore, they do not drift away from their nuclei, making it harder for the stationary electrons to be dislodged from those nuclei upon colliding against the REPT.

Van Allen was wrong when he proposed that charged particles like electrons and protons are trapped in Van Allen belts by geomagnetic fields. He believed charged particles oscillated to and fro within the geomagnetic field. Generally, the movements of gas atoms and molecules in the Van Allen belts are rather random.

Van Allen theorized that the inner Van Allen belt trapped protons, while the outer Van Allen belt trapped electrons. REPT's detection of both electrons and heavier particles in the outer Van Allen belt disproves this theory.

Generally speaking, Van Allen belts have nothing to do with geomagnetic fields, electrons, or protons. Gas atoms and molecules have different capabilities to amass stationary photons at different altitudes, due to their localized gravitational potential energy. When they slam against the REPT, they dissipate dynamic photons to the surroundings. The dynamic photons are captured by the stationary electrons of REPT's high-performance silicone solid-state detector and transform into free electrons, which are recorded as radiation.

Changes in the Van Allen belts hinge on the sun's activities. Gas atoms in the exosphere exchange photons supplied by the sun.

The high radiation in the Van Allen belts can endanger satellites and astronauts. The gas atoms will unload their stockpiles of stationary photons on any object that collides with them. But Van Allen was wrong when he believed that charged particles were trapped in the Van Allen belts due to geomagnetic fields.

# Hurricane

A vast pool of warm water in the ocean is the first prerequisite for the formation of a hurricane. The air above the warm ocean must also be high in humidity. Warm vapor from the warm ocean rises, and some of it condenses into clouds. Through convection, these warm-vapor clouds must attract clouds from the surroundings to hover above the hottest spot on the warm ocean. That is where convection is the strongest, and will likely form the eye of a hurricane.

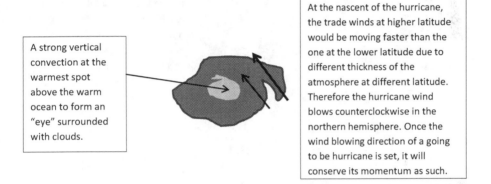

A strong vertical convection at the warmest spot above the warm ocean to form an "eye" surrounded with clouds.

At the nascent of the hurricane, the trade winds at higher latitude would be moving faster than the one at the lower latitude due to different thickness of the atmosphere at different latitude. Therefore the hurricane wind blows counterclockwise in the northern hemisphere. Once the wind blowing direction of a going to be hurricane is set, it will conserve its momentum as such.

Figure 23 – Tropical storm's counterclockwise direction in the northern hemisphere

Every part of a hurricane must support convection to sustain the storm, namely the warm ocean, humid air, and warm clouds. The clouds trap a lot of heat from the sun to fuel further convection. Condensation in the rain clouds in the developing eye dissipates tremendous heat, fuelling further convection. Some heat is retained in the clouds. During the day, the sun constantly energizes the hurricane. The dense clouds trap sufficient heat to fuel convection overnight.

Scientists always underestimate the heat that dissipates when the rain clouds condense to raindrops. Such condensation heat does not just warm up the clouds but also the air trapped between the warm ocean and the clouds. Such dissipated condensation heat may also warm the top portion of the ocean.

Generally, wind blows counterclockwise in the Northern Hemisphere and clockwise in the Southern Hemisphere. Scientists claim that the way tropical storms spin is due to the Coriolis effect, but we think otherwise.

The way tropical storms spin is due to two major factors. The first factor is the trade winds, which blow from east to west on a regular basis. The second factor is the different thickness of the atmosphere at different latitudes. The atmosphere tapers from thickest at the equator to thinnest at the poles. Wind is generally faster at higher latitudes.

A strong vertical convection above the warmest spot on the warm ocean to form an "eye" surrounded with clouds.

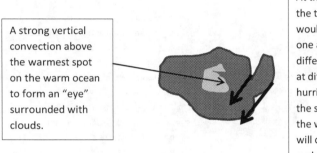

At the nascent of the hurricane, the trade winds at higher latitude would be moving faster than the one at the lower latitude due to different thickness of atmosphere at different latitude. Therefore hurricane wind blows clockwise in the southern hemisphere. Once the wind blowing direction is set, it will conserve its momentum as such.

Figure 24 – Tropical storm's clockwise direction in the southern hemisphere

Figure 23 shows a nascent hurricane spinning counterclockwise in the northern hemisphere. Figure 24 shows a nascent hurricane spinning clockwise in the southern hemisphere. Strong convection encourages air from the surroundings to surge into the storm. This is what the trade winds do, and they may bring in clouds to surround the eye. A nascent hurricane spins as it does because the wind moves slightly faster at higher latitudes, and the atmosphere at higher latitudes is slightly thinner than the atmosphere at lower latitudes.

Trade winds spin around the eye, and the movement of the clouds the trade winds draw in show us the way the hurricane is spinning. The Coriolis effect may seem to answer why hurricanes spin as they do in the Northern Hemisphere, but it may not explain why tropical storms spin as they do in the Southern Hemisphere without ambiguity.

Once a hurricane has gained strength by heightening convection at the eye, the suction of air towards the eye is tremendously strong. This triggers extremely strong winds that move towards the eye more powerfully than the trade winds do. An extremely fast, swirling wind forms around the eyewall of the hurricane, spinning much faster than the wind at the fringe of the storm.

Figure 25 shows the ways convection takes place around a hurricane. Notice that cold air above the hurricane descends to meet with warm air from the eye and is dispersed. This causes the clouds to dispersed from the centre and form the eye.

It would be naive to believe that cold air high in the troposphere sinks through the eye all the way to the bottom of the hurricane, just above the warm ocean. If this were to happen, it would kill off the convection from the bottom and dramatically weaken the hurricane. In reality, this would never happen.

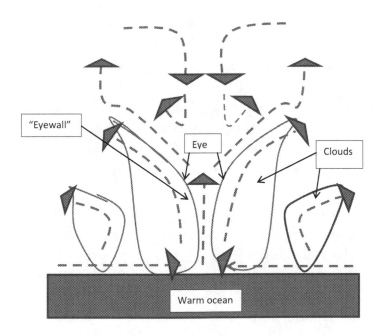

Figure 25 – Convections around a hurricane

# Effects of Global Cooling on Climate Change

Our solar system is located at the fringe of the Milky Way, which is shifting away from the centre of the universe. The earth will experience slowdown as more of its kinetic energy is transformed to universal gravitational potential energy.

As this happens, all atoms on earth will increase their universal gravitational potential energy too. What are the effects of this shift on the atoms? Their nucleons will have an improved capability to stockpile stationary photons over time. So, for example, there would be complete darkness in a remote area on a moonless night, because all earth's atoms would be hungry for more photons.

During the day, atoms are still hungry for stationary photons. But due to higher temperatures, they undergo more intense photon exchange, stockpiling more photons and turning them into stationary photons. At the same time, they release more stationary photons to their surroundings as dynamic photons. This will likely cause record high temperatures in the summer. During the winter, the temperature will drop to extremely low levels because atoms will have an increased appetite for stationary photons. They will not undergo as intense an exchange of photons with their surroundings.

Whether in summer or winter, air molecules have higher mass per time. They have higher inertia and momentum too, as a result of improving their appetite for stationary photons. We can expect winds to blow stronger and faster. This improves their capability to snowball an excessive load of clouds that stand in their paths.

This in turn will have a severe impact on the distribution of rainfall. More erratic rainfall will result in some places experiencing prolonged severe drought. In other places, excessive rain will cause serious floods, because too much rainfall in short time will overwhelm the existing drainage system. In the winter, too much snowfall clogging the highways will dangerous to traffic and will causes roofs to cave in. There may be severe floods when the snow melts.

The earth is not experiencing global warming. This is shown by the record cold temperatures in the winter. If global warming were occurring, we would have moderate winters with milder temperatures. This is not what has happened. Summer is scorching hot and winter is severely cold.

Erratic rainfall is a sign that global cooling is already at the doorstep. We must brace for global cooling before its effects become more pronounced. Steps like planting more trees may help moderate severe weather conditions.

## Geysers, Volcanoes, and Earthquakes

A geyser is a hot spring that spurts a jet of water and steam into the air intermittently. Steam, ash, lava are spewed out during volcanic eruption. Earthquakes share similarities with geysers and volcanoes in that hot steam is released from cracks on earth's crust, especially at the epicentre. For example, kayakers found a rapid stream of bubbles in the ocean near the epicentre after a magnitude 7.8 earthquake rocked Kaikoura, New Zealand, on 14 November 2016.

### Geysers

Geologists believe that surface water seeps underneath the earth's crust and eventually reaches magma, which heats the water into superheated steam. That water eventually channels into the geyser's plumbing system and spurts out as a mixture of water and steam. Liquids and gases flow from high pressure to low pressure regions. Since the pressure at hundreds of meters underneath the earth's crust, where the water heats, is extremely high, it is impossible for surface water to seep into the geyser's plumbing system before reaching the magma.

Discharged water from geysers has a high content of sodium chloride. We suspect that it may be contaminated with seawater from a distant ocean.

### Volcanoes

An erupting volcano is like an exploding pressure cooker. Trapped steam in the magma chamber builds up pressure, enabling deadly eruptions to take place. Again, low-pressure surface water could never get into a high-pressure magma chamber. The geological data derived from collected ash shows that all volcanic ash has high sodium chloride content. Undoubtedly, the water trapped magma chambers originates from the ocean.

Moving magma could readily drag seawater into the mantle along the mid-ocean ridge. The seawater would mix with the magma that flows on the top portion of outer mantle. just below the earth's crust. Since the earth's crust is not homogeneous, a portion could be weak enough to crumble into a magma chamber under the pressure of superheated seawater. The scattering of these points of weakness accounts for why volcanoes are scattered all over earth, not limited to the boundaries of plate tectonics.

Before a volcanic eruption takes place, the earth around the volcano rattles violently, signalling the built-up pressure in the magma chamber. A crater forms on top of the earth's crust. If somehow the magma chamber is clogged by a remnant of lava, then pressure builds again and leads to another volcanic eruption.

## Earthquakes

Large quantities of superheated seawater and steam are trapped between the outer mantle of the earth and its surface bedrock. This exerts tremendous upward pressure against the bedrock. The build-up of steam continues until the bedrock snaps at its weakest spots. This sudden snapping of the bedrock is an earthquake.

A full moon can trigger an earthquake by tugging on the earth's crust. This further weakens the bedrock.

After a major earthquake, the fragments of bedrock re-adjust after most of the steam that supported their weight has gone. The tremors that result from this re-balancing are called *aftershocks*.

Large hydrodams and mining can trigger earthquakes, mainly because those activities can tilt the bedrock underneath and facilitate the trapping of steam.

Some areas will never experience any earthquakes at all, because the bedrock underneath them cannot trap steam.

An earthquake's magnitude depends on the thickness of the bedrock, the strength of the bedrock, the size of the bedrock, and the volume of trapped steam. The depth of the hypocentre determines the level of destruction that occurs. Where the bedrock is deep, the destructiveness is muffled by a thick layer of soil structures. If the bedrock is shallow, there is no massive weight of soil structures to counter the build-up of pressurized steam.

China's Xinjiang is an area constantly rattling with earthquakes. This implies a unique structure in its bedrock. After it snaps, the cracked bedrock quickly seals again, facilitating the trapping of more steam. Of course, a devastating earthquake would take a long time to build up.

The Japanese in the Kanto region fear that a mega earthquake is brewing, since there has been a long respite. If sufficient steam to trigger an earthquake has built up beneath Kanto, constant tremors should be detected. But no one knows for sure if this is happening. Maybe current seismographs are not sensitive enough to detect such tremors.

Some creatures have been noted to have uncanny capabilities to predict earthquakes before seismographs do. Maybe these creatures can lend a hand in alerting us to imminent earthquakes.

The epicentre of an earthquake's destructiveness is where the hypocentre is located. The readings picked up by the seismograph closest to the epicentre are the most accurate, because the earth is not homogenous. Transmitted sound waves vary from place to place. In addition, some energy dissipates during transmission.

# CHAPTER 3

# ELECTROMAGNETIC WAVES

## Electromagnetic Waves – Dynamic Photons per Volume per Time

Electromagnetic waves, or EMWs, are believed to have dual characteristics: they behave like particles in some instances and like waves in others. Do EMWs really have dual characteristics?

It is naive to believe that all EMWs have dual properties simply because we lack better understanding of the forms of EMWs. Quantum scientists treat EMWs as if EMWs know when to behave as particles and to behave as waves. How could EMWs "know" this?

In addition, the assumption that an EMW consists of an oscillating electrical field and magnetic field that are perpendicular to one another is also incorrect. Only charged particles possess electrical fields. Only magnetic objects are surrounded with magnetic fields. Scientists claim that for EMWs, at any time the electrical field or the magnetic field can vanish, implying that matter itself has vanished. To account for this, quantum scientists claim that EMWs are massless. How can EMWs have momentum if they are massless?

Let's assume that all EMWs consist of dynamic photons per volume per time. A higher-intensity EMW has a higher number of dynamic photons per volume per time. That assumption enables us to explain how different EMWs are produced by different devices, such as light bulbs and X-ray machine. We can thus understand their unique characteristics and effects without ambiguity.

When an electrical current passes through the filament of a light bulb, the bulb glows. Dynamic photons are dissipated from the red-hot filament, clearly proving that free electrons are saturated with stationary photons. The light emitted from the bulb causes the vanes of a radiometer to turn, clearly proving that light consists of particles called *photons*. Dynamic photons captured by the vanes cause a change in momentum in the vanes, which causes them to turn.

When a solar panel is under direct sunlight, it produced an electrical current: free electrons. The selenium atom has a rather large orbital. Its stationary electrons easily transform to free electrons after they have gained sufficient stationary photons from the surroundings, especially the sunlight. We are not sure if the unique mosaic of magnetic field of the unique nucleus structure of the selenium atom also plays an important role in helping its stationary electrons to amass dynamic photons from the surroundings. The superposition of an elastic magnetic field with both its stationary electrons and its unique nucleus structure would effectively encourage dynamic photons from the surroundings to adhere to those stationary electrons before transforming them to free electrons.

Maybe the flexing of the unique nucleus structure of the selenium atom also readily transforms stationary electrons to free electrons. Its flexure catapults some of the stationary electrons it has just gained to become free electrons, with the help of a sudden enlargement of the orbital followed by a sudden shrinkage in orbital.

When a solar panel is exposed to the dynamic photons of sunlight, a large number of free electrons are produced. This clearly proves that sunlight consists of dynamic photons per volume per time, not waves. Most of the stationary photons that saturate those free electrons originate from the sun.

Gamma rays emitted from the decaying nucleus of a radioactive atom can also cause the vanes of a radiometer to turn, but at a much faster rate than is caused by a lit bulb, indicating a more dramatic change in momentum. The decaying nucleus spawns a smaller, simpler daughter nucleus whose nucleons require less angular momentum to maintain the cohesiveness of its unique nucleus structure. Excess dynamic photons are released from the restructured nucleus. The dissipating dynamic photons are like fast-moving bullets that embed into the vanes of the radiometer with strong momentum.

If a gamma ray were a wave, then it could not cause the vanes of a radiometer to turn. Massless waves cannot change momentum. Fast-moving photons possess momentum and can cause a dramatic change in momentum to take place in the vanes.

Gamma rays have very high penetrative capability because they consist of a very high number of dynamic photons per volume per time. In fact, among all forms of EMWs, gamma rays have the highest saturation of dynamic photons per volume per time. Gamma rays can penetrate a steel plate easily. Though many dynamic photons are absorbed, a significant number of dynamic photons pass through the steel without interacting with its atoms.

X-ray machines, radio transmission stations, and microwave ovens are all powered by free electrons. The free electrons are saturated with stationary photons. X-ray radiation, radio waves, and microwaves that derive from free electrons are nothing but dynamic photons per volume per time. X-rays have the highest saturation of dynamic photons per volume per time of the three. Radio waves have the lowest.

## The Power of EMWs Is Governed by their Intensity

Quantum scientists believe that various forms of EMWs have different powers because EMWs have different frequencies, as if all of them oscillate like mechanical waves. Scientists also presume all electromagnetic waves have a similar velocity, which is the speed of light. Since frequency times the wave length equals the travelling speed, they reckon that if the frequency is extremely high then the wavelength must be very short or vice versa. Gamma rays and X-rays are very powerful electromagnetic waves believed to have very high frequencies and extremely short wavelengths. Are these assumptions sensible? How do we evaluate the momentum of different EMWs?

A radiometer is a gadget that can detect the presence of EMWs. Based on the speed of the rotating vanes of a radiometer, we can judge the power of a particular EMW. The vanes rotate very fast when subjected to X-rays, and rotate slower when exposed to light emitted from a light bulb.

Every vane on a radiometer has two distinct surfaces: one side shiny, the other blackened. The blackened surface is good at absorbing an EMW if it gains momentum after being exposed to the EMW. It takes the lead in rotating the vanes, which determines their rotational direction.

How the vanes of a radiometer are made to rotate when exposed to an EMW source can be explained easily if we think of the EMW as dissipating dynamic photons with a particular saturation per volume per time.

The photon is a particle, and dynamic photons possess momentum. Dissipating photons are absorbed by the blackened side of a vane. These dynamic photons are transformed to stationary photons, especially those that adhere to nucleons in the radiometer. But they somehow conserve part of their angular momentum. The nucleons thrust in and out while remaining in the unique nucleus structure. This translates to lateral momentum that moves the vanes. Absorption of dynamic photons enables the vane to rotate because the collective change in momentum is significant, though the momentum of a single photon is negligible.

Stronger EMWs, like gamma rays and X-rays, have a high intensity of dynamic photons per volume per time. Generally the vanes of a radiometer turn slower when it is exposed to gamma rays as compared to X-rays, even though X-rays consist of fewer dynamic photons per volume per time than gamma rays. This is because an outburst of gamma rays will only take place when a nucleus is restructuring. The dissipation of photons from an X-ray tube is more consistent, continuing for as long as the X-ray machine is operating.

Each radioisotope emits a different intensity of dynamic photons per volume per time, regardless of the type(s) of emission it has undergone. If the dissipation of dynamic photons is very high intensity, then the radioisotope is also said to have undergone gamma emission. Otherwise, the rather weak dissipation of dynamic photons is recognized as dissipated heat. Dynamic photons are dissipated from a restructured nucleus regardless of the type of emissions.

Furthermore, the radiation emitted from different radioisotopes is not consistent or uniform, because different radioisotopes have different half-lives. The actual occurrence of radioactive decay is also a somewhat random process. Therefore a radiometer is not a good tool for comparing the intensity of gamma rays and the X-rays. But the rotating vanes of a radiometer do prove that both rays consist of dynamic photons per volume per time with different intensity.

In another example, focusing sunlight on the head of a match with the aid of a magnifying glass ignites the match. On the other hand, a match in direct sunlight remains pretty much intact. This clearly proves the constricted sunlight, after it has passed through a convex lens where dynamic photons are saturated in a small space, is more powerful. The highly saturated sunlight on the head of a match can initiate a chemical reaction by saturating both the stationary electrons and the nucleons in a mixture of chemical compounds. Stationary photons in turn enlarge the orbital size. The nucleons in the unique nucleus structure flex in certain orientation and configuration, re-orienting its orbital. This creates an opportunity for the orbital to be receptive to sharing electrons with adjacent atoms. Then a chemical reaction takes place.

Different materials need different saturations of stationary photons per volume per time to be reactive. Therefore, different materials are reactive at different temperatures. Some materials ignite quickly under focused sunlight, while others need longer exposure.

The presence of a higher saturation of dynamic photons per volume per time enables an atom to undergo a more intense exchange of dynamic photons with its surroundings. A nucleon that has absorbed more dynamic photons than it dissipates increases its angular momentum dramatically.

Nucleons with strengthened angular momentum experience a re-orientation of their unique nucleus structure. The unique nucleus structure flexes while still maintaining the uniqueness of its structure. The re-orientation leads to modification of the orbital orientation and configuration. In addition, the denser unique nucleus structure allows its stationary electrons to drift further away from its nucleus, causing enlargement of its orbital size.

Stationary electrons also undergo more intense exchange of photons with the surroundings in the presence of focused sunlight. Stationary electrons that have gained more dynamic photons than they have dissipated are enabled to drift further from the nucleus. I have coined the word *bondbital* to apply to this situation. *Bondbital* is a combination of the words "bond" and "orbital", with the "or" removed. A bondbital is an orbital that can share electrons with like orbitals. Sharing of electrons takes place among bondbitals to form a new bond between reactants. If this happens, then a chemical reaction is said to have taken place.

Thus the presence of sunlight, especially focused sunlight, can initiate chemical reactions. The power of EMWs hinges on the intensity of their dynamic photons per volume per time.

In another example, the colour of the flame of a Bunsen burner changes from yellowish to light blue in colour when the air aperture of the Bunsen burner is opened. When more air is allowed to mix with methane gas, the methane molecule combusts more efficiently. More methane molecules burned per unit of time emits more dissipating photons in the bluish flame than in the yellowish flame. If one places a piece of asbestos above the yellowish flame, the flame leaves black stains because of the presence of incompletely combusted carbon molecules. The yellowish flame is cooler than the bluish flame because fewer methane molecules are actually burning and thus fewer photons are released. The bluish flame contains more dissipating dynamic photons per volume per time than the yellowish flame; therefore different colours possess different saturations of dynamic photons per volume per time. Again, this example suggests that the power of EMWs hinges on their intensity of dynamic photons per volume per time.

Generally, it is believed that the speed of a photon is much faster than the speed of a stationary electron. Is this assumption correct?

We know a stationary electron must gain sufficient photons before it transforms to a free electron. A stationary electron constantly exchanges photons with its surroundings. Dynamic photons from the surroundings adhere to the stationary electron, and stationary photons from the stationary electron are released as dynamic photons. A photon travels much faster than a stationary electron; otherwise photons would fail to catch up with the stationary electron. Free electrons move faster than stationary electrons, but are considered much slower than photons.

A photon is very light, and its energy content is very low. An electron is more massive and its energy content is much higher. However, the energy required to overcome the inertia of photons and electrons is not directly proportional to their mass The inertia energy per mass required by electron is proportionately much higher. Therefore, a dynamic photon possesses much more kinetic energy per mass. A dynamic photon possesses higher speed than a stationary electron, since the photon has a higher percentage of its energy in kinetic energy. The electron has a higher percentage of its total energy in inertia energy, and less in other energies such as kinetic energy.

Inertia energy of photon/Mass of photon << Inertia energy of electron/Mass of electron

Therefore

Kinetic energy $_{photon}$/Total Energy $_{photon}$ > Kinetic Energy $_{electron}$/Total Energy $_{electron}$

Thus

Speed $_{photon}$ >> Speed $_{electron}$

Free electrons can drift from atom to atom because they have higher speeds than stationary electrons. Free electrons leverage the additional energy provided by the stationary photons they have attracted. Less energy is required to overcome a photon's inertia. The photon's greater kinetic energy helps move the free electron faster. The more stationary photons on a free electron, the greater the boost to the electron's speed.

Not every free electron possesses a similar speed. It depends on their load of stationary photons. Free electrons that are more saturated with stationary photons have higher speeds.

Free electrons at lower voltage can travel a shorter distance than free electrons at higher voltage. For this reason, power transmission over distance requires high voltage. Free electrons at high voltage have a higher saturation of stationary photons and can travel at higher speeds over greater distances. Free electrons are likely to shed some of their stockpile of stationary photons during the transmission process. So it is not wise to have power plants located far away from consumers. It is always good to have power plants scattered around, to cut down on power wastage.

When free electrons flow through a wire, the magnetic field around the wire becomes stronger. The more free electrons that flow, the stronger the wire's magnetic field becomes. Every individual free electron acts like a tiny magnet. A cluster of them superpose to create a stronger magnetic field. A free electron saturated with stationary photons has an even stronger magnetic field, since stationary photons superpose and create a strong magnetic field themselves, surrounding the free electron. A strong magnetic field helps free electrons retain stationary photons.

Let's say a dynamic photon adheres to a stationary electron. The speed of a dynamic photon is much faster than that of a stationary electron; therefore, a portion of its kinetic energy is preserved in magnetic energy by boosting its intrinsic spin. In addition, a portion of the kinetic energy of a photon boosts the stationary electron's speed, which is kept in the form of angular momentum.

A stationary electron also possesses intrinsic spin. A stationary photon perched on a stationary electron is like a compressed spring. Its newly boosted magnetic field merges with the magnetic field of the stationary electron to form an elastic magnetic field. A big portion of the dynamic photon's kinetic energy is preserved in other forms when it perches on a stationary electron, which is drifting more slowly than the speed of light.

Stationary electrons revolve around nuclei. Therefore they may not have full bearing in conserving the momentum of the atoms in the vanes of a radiometer rotating around a pivot. More massive stationary electrons, saturated with stationary photons, agitate the nucleus of an atom as they revolve around the nucleus, causing the nucleus to wobble. This makes it easier for the agitated vanes to rotate, since the vanes gain higher momentum. The vanes of a radiometer rotate mainly due to the absorption of stationary photons by their nuclei. Embedding stationary photons in a nucleon is more effective in elevating the momentum of the vanes than agitating the vanes with more massive stationary electrons.

A dynamic photon that adheres to a nucleon transforms a big portion of its kinetic energy to intrinsic spin, which subsequently transforms to intrinsic magnetic energy because those nucleons are somewhat stationary.

Dynamic photons adhere to both stationary electrons and nucleons. These photons add some of their kinetic energy to help boost the momentum of the stationary electron or nucleon. Stationary photons that adhere to nucleons keep a bigger portion of their kinetic energy to boost the angular momentum of the nucleon, which in turn boosts the momentum of the atoms of the vanes of a radiometer. In contrast, the dynamism of the more massive stationary electrons that revolve around nuclei does not directly boost the momentum of the atoms of the vanes. The vanes of a radiometer thus rotate more strongly, proving that a portion of dynamic photons' momentum is conserved as the momentum of atoms in the vanes, and eventually translates to rotational motion of the vanes.

Some stationary photons that adhere to nucleons or stationary electrons are transformed back to dynamic photons again and scatter to the surroundings. The vanes might experience reduction in momentum after the stationary photons transform.

How fast the vanes rotate depends on their net gain in momentum. Momentum gained minus momentum lost is equal to the net gain in momentum. This directly proves that photons are tiny particles that possess mass, and dynamic photons possess momentum. Arresting a dynamic photon causes a change in momentum.

X-rays makes a radiometer's vanes rotate faster than sunlight does. Since an X-ray is a more intense EMW with a higher saturation of dynamic photons per volume per time than ordinary sunlight, X-rays cause the radiometer vanes to absorb more dynamic photons per time than they dissipate to the surroundings. The vanes gain more momentum, enabling them to rotate at a faster speed.

The belief that EMWs consist of oscillating magnetic and electrical fields that are perpendicular to one another cannot explain how the vanes of a radiometer gain momentum after being exposed to a source of EMWs.

Sunlight can be focused by a magnifying glass, and focused sunlight can ignite a piece of newspaper. This directly proves that focused light is more powerful than ordinary sunlight. The dynamic photons passing through a magnifying glass are constricted to a small area, enabling the focused light to initiate a chemical reaction. If sunlight were a wave, how could a wave passing through a magnifying glass become more powerful? Ripple tank experiments show that waves before and after passing through a lens are exactly the same in terms of wavelength. This phenomenon strongly suggests that all EMWs are dynamic photons per volume per time.

It is not true that a radio wave has very long wavelength and tall amplitude. Even though a radio's antenna is rather short, the antenna can tune in to a specific channel. A short antenna could hardly intercept the entire radio wave. This strongly suggests that a radio wave is dynamic photons per volume per time. When the specific saturation of dynamic photons per volume per time of a radio wave is intercepted by an antenna, a microcurrent is produced by the adherence of those dynamic photons to stationary electrons, which are then transformed to free electrons.

Gamma rays can penetrate even a thick steel plate. Despite the sizeable number of dynamic photons that have been absorbed by the nuclei of iron atoms, there are still many dynamic photons are that succeed in passing through the plate. Therefore, gamma rays are very penetrative. In addition, gamma rays turn the vanes of a radiometer much faster than ordinary light does, which hints that a gamma ray consists of a higher intensity of dynamic photons per volume per time.

All forms of EMWs demonstrate characteristics that support the notion that EMWs consist of dynamic photons per volume per time with different intensities. Wave theory has failed miserably in accounting for these characteristics. Gamma rays, with the highest saturation of dynamic photons per volume per time, are the most powerful form of EMWs. The power of an EMW hinges on its saturation per volume per time, rather than frequency or wavelength.

## Why Photons Can Travel Intergalactically

Every photon is believed to be a tiny magnet. According to the Biot–Savart law, the emergence of an induced magnetic field prevents a magnet from making a displacement. For example, let's say a magnet is intended to move to the right. An induced magnetic field emerges, sandwiched between the existing magnetic field of a magnet that attempts to prevent such a magnet from displacing it, as shown in Figure 26. The newly induced magnetic field that opposes an attempt to displace the magnet used to be part of the existing magnetic field of the magnet before the displacement.

Once the displacement has taken place on the magnet, existing magnetic fluxes, especially at the fringe of the existing magnetic field of the magnet, are transformed to opposing north and attractive north, which attempt to resist the displacing magnet, as shown in Figure 26.

The reasons for the formations of opposing north and attractive north are still pretty much an unresolved mystery. Maybe the interaction between the north magnetic fluxes, especially at the fringe of the magnetic field, leads to the formation of opposing north magnetic fluxes. The nudging flips the north magnetic fluxes at the fringe over to form opposing north, while the south magnetic fluxes that are dragged at its fringe transform to attractive north. Maybe when the south magnetic fluxes at the fringe of the south polarity of the magnetic field have been dragged, they flip before transforming to attractive north magnetic fluxes.

This shows that magnetic north and south are related to one another. Attractive north magnetic fluxes used to be part of south magnetic fluxes. But in what ways they are related is still perplexing.

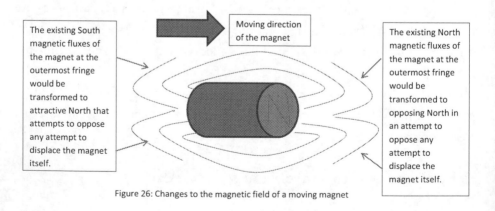

Figure 26: Changes to the magnetic field of a moving magnet

We use the term *attractive north magnetic fluxes* to explain that such magnetic fluxes are attracted to south magnetic fluxes as if they are north magnetic fluxes. They do not really have anything to do with north magnetic fluxes, since they remain as part of the south magnetic fluxes.

The formations of opposing north and attractive north magnetic fluxes at the fringe of a magnetic field depend on the physical action (nudge against or drag away) of inner magnetic fluxes against the fluxes at the fringe of the magnetic field. Once displacement stops, opposing north and attractive north magnetic fluxes quickly revert to north and south magnetic fluxes, as is shown in the status quo condition in Figure 27.

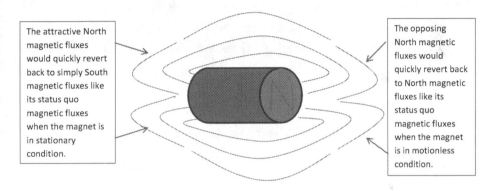

The attractive North magnetic fluxes would quickly revert back to simply South magnetic fluxes like its status quo magnetic fluxes when the magnet is in stationary condition.

The opposing North magnetic fluxes would quickly revert back to North magnetic fluxes like its status quo magnetic fluxes when the magnet is in motionless condition.

Figure 27: Changes to the magnetic field of a magnet as soon as it stops moving

Any attempt to initiate movement to a permanent magnet leads to changes in its magnetic fluxes, especially the ones at the outermost fringe, where newly transformed magnetic fluxes attempt to resist movement to the magnet. Understanding this phenomenon enables us to comprehend the reasons why dynamic photons can travel intergalactically.

The Biot–Savart law is true because the current induced by a moving magnet in a solenoid takes place depending on the manner of movement of the magnet itself, as dictated by Lenz's law. Lenz's law states that the polarity of the induced electromagnetic field (EMF) produces a current that creates an opposing magnetic flux. Therefore, at least based on the induced current, we are certain that the Biot–Savart law is correct when the magnetic fluxes at the fringe of the magnet are flipped over, opposing any changes to its movement.

A dynamic photon is like a tiny magnet. Its advancement causes the induced magnetic fluxes at its fringe to sandwich it, resisting its movement forward. Let's say the magnetic orientation of a photon is exactly like the magnet in the previous example. This photon tries to thrust forward with acting force F1 (its initial angular momentum), as shown in Figure 28. As it thrusts forward, the opposing north at the fringe emerges to oppose its advancement. As the photon continues to thrust forward, more inner north magnetic fluxes transform to stronger, opposing north magnetic fluxes. The emergence of an opposing north magnetic field helps form an elastic magnetic flux with the existing north magnetic fluxes of the photon.

The photon will then transform more kinetic energy to magnetic energy. The elastic magnetic field will become more compact and elastic. The photon will experience slowdown, as shown in Figure 29.

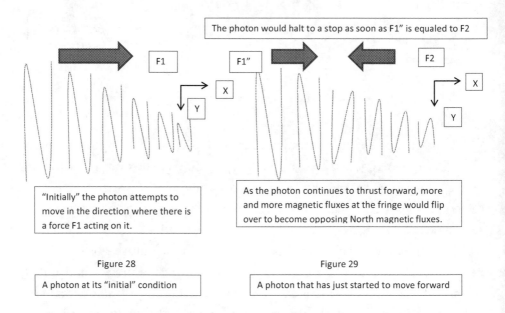

The photon would halt to a stop as soon as F1″ is equaled to F2

F1     F1″     F2     X     Y     X     Y

"Initially" the photon attempts to move in the direction where there is a force F1 acting on it.

As the photon continues to thrust forward, more and more magnetic fluxes at the fringe would flip over to become opposing North magnetic fluxes.

Figure 28

A photon at its "initial" condition

Figure 29

A photon that has just started to move forward

As soon as the elastic magnetic fields collapse, the north and the opposing north magnetic fields vanish. A new acting force emerges. (We believe that one energy can transform to another energy. Here, magnetic energy transforms to an acting force against the photon in a way that will increase its kinetic energy eventually.) The new acting force diverts the photon into a new direction. Now the photon is acting by a stronger force F3, as shown in Figure 30.

When the photon thrusts forward in a new direction, the photon increases its intrinsic spin and transforms more of its kinetic energy to magnetic energy, as shown in Figure 31. Acting force F3 weakens to acting force F1. The photon is back to its status quo again (Figure 28), except its direction of motion has changed. The process illustrated in Figure 29 will again take place in this new direction, and a similar sequence of processes will continue endlessly. Overall, the photon moves forward in a tight spiral.

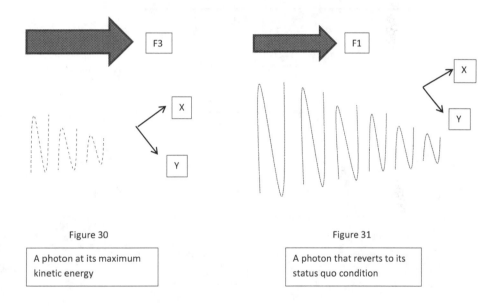

Figure 30

A photon at its maximum kinetic energy

Figure 31

A photon that reverts to its status quo condition

Figure 32 shows the movement of a photon on a Z-plane. The magnetic fluxes of the photon along the Z-plane at location (1) are in neutral condition. All magnetic fluxes are in status quo condition.

As the photon spirals upward, the north magnetic fluxes at the fringe flip over to build up elastic magnetic fluxes, until the forces of the north magnetic fluxes and opposing north magnetic fluxes are almost equal at location (2). The elastic magnetic fluxes collapse. Subsequently, the photon is pushed downward along the Z-plane.

This acting force boosts the kinetic energy of the photon, as if magnetic energy has already been transformed to kinetic energy. As the photon drifts forward at a faster velocity, its intrinsic spin is hastened. This transform more kinetic energy to boost the magnetic field surrounding the photon. When the photon reaches location (3) in Figure 32, the photon magnetic field along the Z-plane is back to its status quo condition.

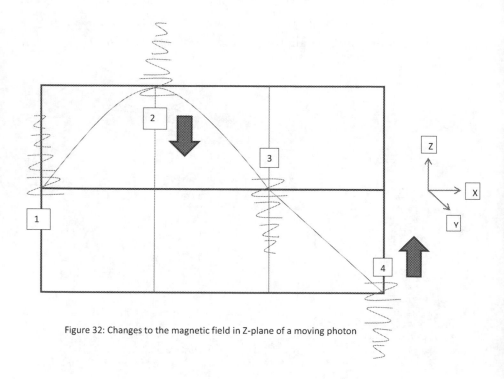

Figure 32: Changes to the magnetic field in Z-plane of a moving photon

As the photon continues to thrust forward, the newly emerged, opposing north magnetic fluxes build up gradually in a way that slows the photon again. This causes the photon to hasten its intrinsic spin while transforming more of its kinetic energy to boost its magnetic energy. The strengthening magnetic field gradually transforms more magnetic fluxes, especially, at the fringe of opposing north magnetic fluxes. This builds up the elastic magnetic field between the magnetic field of the photon and its opposing north magnetic fluxes. The strengthening of the elastic magnetic field enables the deflection the photon when the photon is at location (4). Again, the process repeats indefinitely as the photon spirals along a straight path on the Z-plane.

If we understand the changes in terms of the north magnetic fluxes of a moving photon along the Z-plane, we can apply the same context to understanding the movement of a photon along a Y-plane. But location (1) on the Y-plane (using Figure 32 with the same location numbering, but renaming the plane) also refers to location (4) on the Z-plane when you view the moving photon along the Y-plane. A specific portion of the north magnetic fluxes of the photon are always parallel to its axes. For example, when the photon is at location (1) in the Y-plane, the north magnetic field in question is oriented along the Y-axis.

Though photons are extremely light with very small inertia, dynamic photons can travel intergalactically because they are capable of recycling a portion of their kinetic energy and magnetic energy, if there is an acting force continuously poking at them.

Light travels in a straight line, but individual dynamic photons can't possibly move forward in a straight line. There is always an opposing north magnetic field emerging to cause them to move "sideways", in a spiral path. Dynamic photons spiral in a straight line.

This is a theoretical explanation. We can use something like a beam of free electrons that are streaming in a uniform magnetic field to help us see how a dynamic photon can spiral forward in a straight line.

A beam of free electrons spirals in an almost perfect circular path when the free electrons are traveling in an external uniform magnetic field, especially a magnetic field that is perpendicular to the direction of motion of those free electrons. One may argue that the presence of an external magnetic field that is constantly interacting with the magnetic field surrounding the free electrons is not similar to the dynamics experienced by a moving photon. But the focus here is to show how a free electron spirals the way a dynamic photon does. We also strongly believe that free electrons move in spiral paths in the absence of an external magnetic field.

The spiral of a beam of free electrons clearly shows that there is a centripetal force constantly acting against the free electrons. It is a result of interaction between the magnetic fields of the free electrons and the external magnetic field, which is constantly perpendicular to the direction of motion of those electrons. The motion of the free electrons is much more complex than the motion of dynamic photons because the free electrons are also constantly undergoing exchange of photons with their surroundings. The process of exchange may affect their moving loci. But the absorption and dissipation of photons plays second fiddle role in the moulding of their circular path.

In this segment, we will discuss the outcome of an experiment conducted by Dorothy Montgomery, chair of the department of physics of Hollins College, in 1959 to show how moving electrons in a perpendicular uniform magnetic field behaved. (See "Electrons in a Uniform Magnetic Field" at https://youtu.be/gwbRMXYkC3M.)

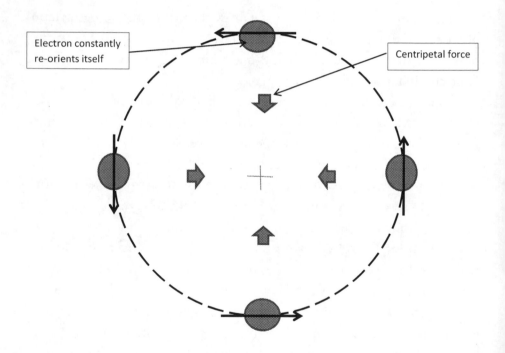

Figure 33: Centripetal force acting on an electron moving in a circular path

A glass bulb with an electron gun inside is snugged between two perpendicular rings, which are wrapped with coils that produce a uniform magnetic field when a current is flowing. The electron gun is positioned upright so that the moving free electrons are perpendicular to the external uniform magnetic field.

Initially the electron gun is switched on while there is no current running through the coils. One sees an upright beam of free electrons in the glass bulb. Then a current is allowed to flow through those coils, allowing the emergence of an external uniform magnetic field. As the external magnetic field is gradually strengthened, the beam of free electrons is bent bit by bit until it forms an almost complete circle. The radius of the circular path shrinks as the external magnetic field is strengthened.

Since the free electrons move in a circular path, this clearly shows that there exists a centripetal force that is constantly acting against the free electrons as a result of the interaction between the external uniform magnetic field and the magnetic field of the free electrons themselves. The orientation of the external magnetic field is fixed; therefore the orientation of the magnetic fields of the free electrons must be constantly re-orienting. An elastic magnetic field builds between the magnetic field of the free electrons and the external magnetic field that is perpendicular to those electrons.

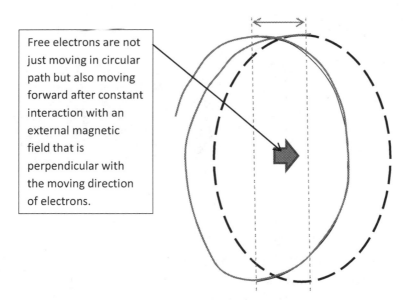

Free electrons are not just moving in circular path but also moving forward after constant interaction with an external magnetic field that is perpendicular with the moving direction of electrons.

Figure 34: An electron propels forward while moving in a circular path

Even if the electron-gun is in an upright position, the free electrons it emits are not just moving in a circular path but also moving slightly forward, as shown in Figure 34. This is mainly because there is an unbalanced force propelling them forward, resulting from the magnetic-magnetic interaction between the external magnetic field and the electrons.

In addition to the centripetal force, there is another acting horizontal force on the free electrons. Though the external magnetic field is fixed and perpendicular to the electron gun, the acting force (a force that results from the coupling of the centripetal force and a horizontal force, shown by dark green arrows) is at an angle due to the interaction between the magnetic field of the free electrons and the external magnetic field, as shown in Figure 35.

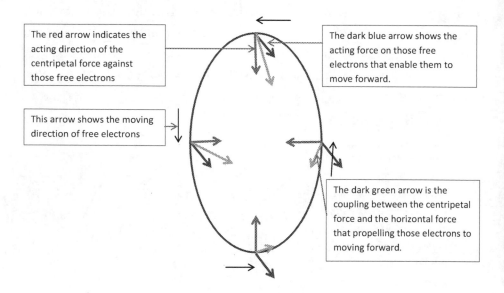

The red arrow indicates the acting direction of the centripetal force against those free electrons

The dark blue arrow shows the acting force on those free electrons that enable them to move forward.

This arrow shows the moving direction of free electrons

The dark green arrow is the coupling between the centripetal force and the horizontal force that propelling those electrons to moving forward.

Figure 35: Acting forces on an electron that moving in a circular path

The existence of an acting force that is constantly propelling those free electrons to move forward is also a result of the interaction between the magnetic field of the electron and the external magnetic field that is constantly perpendicular to the moving free electrons. This clearly shows that the outcome of the interaction between the external magnetic field and the free electrons is not exactly as dictated by Fleming's left-hand rule. An extra acting force is propelling those free electrons to move forward.

The kinetic energy of free electrons cannot interact with the external magnetic field. It must be the interaction between the magnetic field of the free electrons and the external magnetic field that enables them to spiral. Bear in mind that the moving electron also has its own opposing magnetic field that resists its motion, and this opposing magnetic field also interacts with the magnetic field of the electrons.

Professor Montgomery collected a set of data as shown in Table 10. The voltage is the applied voltage on the electron gun, while $B$ is the magnetic strength of the external magnetic field, and $r$ is the radius of the circular path travelled by the free electrons after they have interacted with the external magnetic field.

| Voltage (x1.6xExp-19 J/el ch.) | B (x Exp-22 N/el ch) | r (xExp-2 m) |
| --- | --- | --- |
| 75 | 0.78 | 6.38 |
| 100 | 1.23 | 4.30 |
| 150 | 1.23 | 5.30 |
| 200 | 1.23 | 6.25 |

Table 10: Collected data on an experiment of electrons moving in circular path

An increase in voltage naturally increases the current, which causes the free electrons to travel in a much larger circular path. This is because free electrons produced at a higher voltage are more saturated with stationary photons. They are more massive with higher inertia.

As these free electrons spiral in a circular path, they also undergo an exchange of photons. They may dissipate dynamic photons to the surroundings while absorbing other dynamic photons. This indicates that the mass of individual electrons may not be the same, depending on their stockpiles of stationary photons.

In addition, free electrons produced at different voltages have different stockpiles of stationary photons. Higher voltages produce free electrons with a higher saturation of stationary photons. Though they also have stronger magnetic fields to interact with the external magnetic field, their mass causes them to spiral in a much larger circle.

A stronger external magnetic field interacts with the magnetic field of the dynamic electrons more strongly, creating a stronger elastic magnetic field and producing a stronger centripetal force that acts on the electrons. They travel in shorter free paths, or smaller circular paths.

The data collected in this experiment cannot be used to calculate the mass of an electron, because the free electrons emitted by the electron gun at different applied voltages have different saturations of stationary photons. Higher voltage allows the production of much denser free electrons. Even if the applied voltage on the electron gun remained constant, changes to the external magnetic field's strength influence how well the free electrons maintain their stockpiles of stationary photons. A strong external magnetic field may enhance the electrons' capabilities to retain their stockpiles. A weaker external magnetic field may encourage free electrons to shed more of their stationary photons.

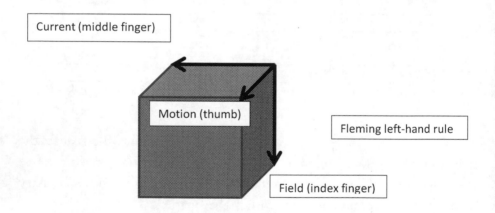

Figure 36: Fleming left-hand rule

We may want to use Fleming's left-hand rule to help us to predict the direction of motion of a charged particle drifting in an external magnetic field, as shown in Figure 36. Fleming's left-hand rule is rather handy and somewhat accurate for determining the direction of motion of the charged particle. The defect is that the new direction of motion of the charged particle is the result of magnetic-magnetic interaction between the charged particle and the external magnetic field, rather than current-magnetic interaction as reckoned by Fleming's left-hand rule.

Based on a beam of free electrons streaming in a fixed external magnetic field, we know with certainty that a resultant force (green arrows in Figure 35) acts on the free electrons at an angle to the electrons, as shown in Figure 35. Fleming's left-hand rule only helps us to determine the acting direction of the centripetal force (red arrows in Figure 35). Fleming's left-hand rule fails to predict the actual resultant force (green arrows) acting on the free electrons. The electrons are not just moving in a circular path, but are also moving forward.

The magnetic field strength among distant stars in intergalactic space may be negligible, since magnetic influence among stars is not far-reaching. Therefore, dynamic photons can travel forward while spiralling as a result of their own magnetic field interacting with their newly emerged opposing magnetic field. Subsequently, they exchange part of their magnetic energy for kinetic energy and propel themselves forward.

Elastic magnetic fields build up between the opposing north and north magnetic fluxes of the photon to some maximum level, then collapse. They transform to an acting force that acts against the photon. Magnetic energy is transformed to kinetic energy and the elastic magnetic field vanishes. The photon increases its kinetic energy by travelling faster in a new direction, under the influence of the new acting force. Its intrinsic spin slows in this transformation to kinetic energy.

As the photon accelerates forward, its magnetic field builds up again long before its magnetic field gradually transforms to the opposing magnetic field. The photon slows down as the elastic magnetic field strengthens further. It slows further as it hastens its intrinsic spin, which transforms more of its kinetic energy to magnetic energy and strengthens the elastic magnetic field. When the elastic magnetic field reaches maximum capacity, it collapses. An acting force emerges. This process continues as the photon spirals forward.

Dynamic photons spiral forward. This hints that their magnetic fields and currents might not be exactly perpendicular to one another, as dictated by Fleming's left-hand rule. We strongly believe the location where opposing north emerges on the photon constantly shifts, and that enables a dynamic photon to spiral forward. In addition, the emergence of a resultant acting force acts at an angle that enables the photon to spiral while travelling in a straight path at a great velocity, $3 \cdot 10^8$ meters per second.

Dynamic photons are able to travel intergalactically because they constantly recycle their kinetic energy and magnetic energy. There is always an external force acting against it. Otherwise, photons, which possess extremely light mass and little energy and momentum, would not be able to travel intergalactically.

We are still unsure how dynamic photons maintain their movement in a straight line while spiralling. They indisputably do, since light travels in a straight line.

We believe there is one particular patch in the opposing magnetic field that surrounds the dynamic photon which emerges as the elastic magnetic field is strengthening. The elastic magnetic field reaches its maximum strength and collapses. Suddenly there emerges an acting force that acts against the photon. It's likely that the portion of the dynamic photon where the patch of opposing magnetic field emerges strengthens before waning. The patch constantly shifts to be parallel to the direction of the acting force. The collapse of a particular patch of elastic magnetic fields not only fosters the centripetal force but also the acting force. We don't know how significant the acting force is.

Let's assume quantum scientists are correct in estimating the speed of electrons at approximately 80 per cent of speed of light. This implies that the stationary photons adhering to stationary electrons have less than 20 per cent of their total energy in magnetic energy. Assume the opposing magnetic field of a dynamic photon is as strong as the magnetic field of the photon itself, before the elastic magnetic field collapses and transforms magnetic energy to kinetic energy. This implies that the opposing magnetic energy could be less than 10 per cent of the photon's total energy.

Even though less than 20 per cent of the photon's energy is magnetic, that percentage is recycled. There exists a constant acting force that repeatedly acts against the photon and enables it to travel intergalactically.

Propulsive spacecraft like rockets may not able to travel intergalactically because oxygen, which is crucial to sustain life, must be utilized to burn fuel and propel the rocket forward. A spacecraft mimicking the way a photon travels, interacting with its own magnetic field and recycling its magnetic energy, would enable humankind to travel beyond the solar system in the future. Even if such a spacecraft can attain only 10 per cent of the speed of light, that would be much faster than conventional rockets. Maybe unidentified flying objects (UFOs) mimic the way photons travel intergalactically.

Surrounding a rocket with a strong magnetic field may shield it from projectiles. A rocket will explode if hit by a projectile, since punctures to the fuel tank, the oxygen tank, or both are disastrous.

## The Photon Is a Negatively Charged Particle

Free electrons are saturated with stationary photons. Free electrons of different potentials have different saturations. Free electrons of higher potential differences have higher saturations of stationary photons.

At higher potential differences, a cathode tube has better "suction" between cathode and anode. Free electrons from a cathode with a higher potential difference have stronger attraction to the anode of a cathode tube. More saturated free electrons with stockpiles of stationary photons are more negatively charged. Thus, there is no doubt that a photon is a negatively charged particle. This discovery strongly suggests that the electron is not homogeneous. It contains a positive quark that those stationary photons can adhere to.

Since the photon is a negatively charged particle, it must be enveloped in a magnetic field with a specific orientation and configuration. The earth is a big magnet with the North Pole at the southern geographic pole and the South Pole at the northern geographic pole. So a photon interacts differently with the different magnetic polarities of the earth.

There are differences in terms of brightness and duration in the seasonal sightings of the aurora near the North Pole as opposed to the one at the South Pole. The aurora near the North Pole is called the *aurora borealis*. The aurora near the South Pole is called the *aurora australis*. The aurora borealis is much brighter than the aurora australis. During the winter, when the saturation of air molecules in the magnetosphere is dense enough to allow the exchange of photons, the exchange is seen as light in the sky. The aurora borealis is seen between September and mid-April. The aurora australis is seen between mid-April and mid-August. The duration of sightings of the aurora borealis is seven and a half months, while the duration of sightings of the aurora australis is only four months.

Auroras occur when geomagnetic strength is sufficient to encourage an intense exchange of photons. At particular altitudes, the effect of localized gravitational potential energy allows maximum exchange of photons to take place. Auroras typically originate in the thermosphere, which is eighty kilometres above sea level.

During a solar flare, the aurora can be seen at much lower latitudes because the greater supply of dynamic photons from the sun encourages a more intense exchange of photons among air molecules at high altitude, even in the presence of a much weaker geomagnetic field.

Auroras are seen much longer in the Arctic than the Antarctic. Auroras in the Arctic are also much brighter than the ones in the Antarctic, implying that dynamic photons, especially from the sun, are more penetrative in the Arctic than in the Antarctic. Maybe the head of a dynamic photon has a north magnetic polarity. Similar magnetic poles repel one another, so the north magnetic polarity of the photon head would be repelled by the north geographic pole of the South Pole. Dynamic photons from the sun are deflected from the South Pole.

NASA has reported that the Antarctic sea ice sheet expanded in 2013 to its maximum size. This scientific discovery supports the notion that the Antarctic is much colder than the Arctic because photons from the sun are more accessible in the Arctic. The sea ice sheet melts more extensively in summer in the Arctic than the one in the Antarctic.

Understanding auroras enables us ascertain that the photon is a negatively charged particle with a north magnetic polarity.

## Sources of EMWs

Among the forms of EMWs, the gamma ray is the most powerful of all, followed by the X-ray. The ultraviolet ray is next, then infrared, visible light, and microwaves. The least powerful of all are radio waves.

## Gamma Rays

Quantum scientists believe the presence of a strong force makes an element remain non-radioactive. The presence of a weak force transforms an element to a radioactive substance which subsequently decays. Why the strong force exists and how it maintains the stability of a nucleus are questions that have not been fully answered. Likewise, why the weak force exists and how it causes the restructuring of a nucleus has not been satisfactorily explained.

The emission of gamma rays by radioactive substances strongly suggests that the nuclei of all radioactive substances are saturated with stationary photons. The weakening angular momentum of nucleons in radioactive substances can no longer resist repulsive forces, especially among protons. This is due to the expansion of the universe. Restructuring of the nucleus inevitably takes place, and excess dynamic photons are dissipated to the surroundings. This is because a simpler and smaller nucleus requires less angular momentum to keep it stable. Sometimes other types of emission also take place, like proton emission, beta emission, alpha emission, neutron emission, and so forth, depending on the type of radioisotope. Dissipation of heat and other forms of emission during the restructuring of radioactive substances is recognized as *radioactivity*.

Gamma rays are highly penetrative. We previously mentioned that they can pierce a thick iron plate. They can also pierce a thick lead plate. This is because gamma rays have the highest saturation of dynamic photons per volume per time. A sizeable number of dynamic photons are absorbed by the nucleons of the lead's atoms when a gamma ray penetrates it. But there are still plenty of dynamic photons that succeed in passing through without interacting with the lead atoms.

It is naive to believe that gamma rays are very penetrative because the wavelength of gamma rays is very short and the frequency very high. Gamma rays can cause the vanes of a radiometer to spin at a very fast rate, suggesting that absorption leads to a notable change in momentum. Photons possess mass, and dynamic photons have the momentum to create this notable change. Massless EMWs could never turn the vanes of a radiometer.

## X-rays

High-voltage direct current (DC) is used to power an X-ray machine. Free electrons of a higher tension power are saturated with more stationary photons. When a cluster of free electrons from the cathode bombard the anode, dynamic photons are released as those free electrons convert to stationary electrons at the anode.

X-rays have strong penetrative power, but less than gamma rays. Dynamic photons of X-rays can cause chemical reactions on a film that has been shielded in a steel case. Photons of X-ray radiation react with silver halides on the film to form an image in silver.

## Ultraviolet, Visible Light, and Infrared

Quantum scientists believe that ultraviolet has a shorter wavelength than visible light, and visible light has a shorter wavelength than infrared. In other words, ultraviolet is the most powerful of the three and infrared is the weakest.

But infrared turns out to be more powerful than visible light. Infrared can heat up a thermometer more effectively than visible light. This causes us to doubt the definition of infrared in quantum mechanics.

Formation of ultraviolet, visible light, and infrared takes place when sunlight passes through a prism. Dynamic photons in the sunlight are dispersed by the prism to form a spectrum. Therefore, it is not completely right to say that ultraviolet has the shortest wavelength and infrared the longest. Ultraviolet has the highest saturation of dynamic photons per volume per time of the three, and visible light has the lowest saturation. Therefore, ultraviolet is the most powerful of the three, and visible light is the least powerful. How come?

We don't know why, but it is true that sunlight passed through a prism disperses to ultraviolet, visible light, and infrared. Different saturations of dynamic photons per volume per time in every colour of a spectrum depend upon how the prism disperses sunlight. Each wavelength of sunlight has a unique saturation of dynamic photons per volume per time. Ultraviolet has the highest saturation of dynamic photons per volume per time, followed by infrared, then visible light. This is how a cluster of dynamic photons in sunlight are deflected by the unique nucleus structure of the molecules of the prism. We reckon that prisms made of different materials display a different spectrum when sunlight is passed through them.

## Microwaves and Radio Waves

Microwaves emitted by a microwave oven have a high intensity of photons per volume per time, capable of cooking food or heating water in a short time. Visible light cannot cook food or boil water.

A radar mounted on a self-driving car exposes people in its vicinity to unnecessarily high radiation, which may cause a steep rise in cancer cases in the foreseeable future. In addition, self-driving cars are a complete failure in congested streets, where too many dynamic photons confuse them. Every self-driving vehicle emits radio waves. Each vehicle's radar wrongly perceives them as reflected radio signals, which indicate obstacles. In the presence of other self-driving vehicles, the vehicle perceives that there are too many obstacles to allow it to move forward. All self-driving cars would come to a halt on a crowded city street.

Radio waves are transmitted from a radio station and intercepted by a radio antenna. The antenna transforms some stationary electrons in the LC or tuned circuit to free electrons. This radio signal is amplified by a transistor, transforming it to a stronger signal current that feeds to a speaker. The speaker then produces meaningful sounds.

Close to a radio station, the radio waves it emits may be rather strong. We don't know if there is a rise in cancer cases among residents living near a radio station. No one knows how safe radio waves and microwaves are.

## Shiny Metal and the Saturation of Stationary Photons on Nucleons

A sheet of shiny metal can act like a mirror because it reflects much of the light that shines on it. Metallic atoms are compact and arranged in an orderly manner, so the light they reflect is coherent and parallel. The unique nucleus structure of metallic atoms is very good at exchanging photons.

The atoms absorb some dynamic photons but re-emit most of the dynamic photons that they have already absorbed. Therefore, most metallic atoms possess a shiny surface, as shown in Figure 38.

If the atomic size of the atoms is not the same, the arrangement of those atoms is not orderly. The surface they create is jagged. Such metallic atoms do not reflect light coherently, as shown in Figure 37.

Light from an object strikes a shiny metallic sheet which acts like a mirror. Dynamic photons are briefly absorbed, especially by the nuclei, then dissipated again, which allows us to see the object in the "mirror" with clarity. Most of the dynamic photons are deflected in this way; only a small number are absorbed for the long term by the metallic nuclei.

This phenomenon proves that a dynamic photon is a particle. When it strikes another object, such as a nucleon, it can be deflected.

The angle of incidence and angle of reflection of light hitting a metallic atom are almost identical. This suggests that the geometrical shape of a metallic nucleus is spheroid. A symmetrical, spherical nucleus fosters a symmetrical, spherical electron shell, ensuring a more coherent arrangement of atoms like the one shown in Figure 38.

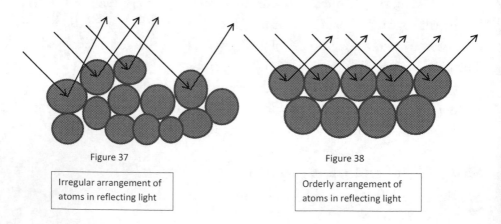

Figure 37

Irregular arrangement of
atoms in reflecting light

Figure 38

Orderly arrangement of
atoms in reflecting light

We used to believe that when light shone on an object, dynamic photons were reflected by the electron shell of the atom. Since stationary electrons revolve continually around the nucleus, they have limited interaction with dynamic photons. Otherwise the angle of incidence of light would not necessarily equal its angle of reflection. The way light is reflected depends on the geometrical shape of the nucleus, not the electron shell. If incoming light is at "normal" to the metallic sheet, the reflected light is also at normal to the mirror.

At ambient temperature, a photon that travels at great speed towards a nucleon before being deflected away does not disintegrate into smaller fragments. This is because a photon is one of the simplest form of particles, and it is enveloped by a magnetic field. The build-up of an elastic magnetic field around the photon cushions the impact with the nucleon, which allows the photon to have a soft landing on the nucleon before being deflected.

Just before the photon touches down on the nucleon, the magnetic field of the photon interacts with the nucleon. The emergence of an elastic magnetic field nudges the photon to slow down. It transforms some of its kinetic energy to hasten its intrinsic spin, which transforms more of its kinetic energy to magnetic energy, allowing the build-up the elastic magnetic field between them. The photon stops moving forward momentarily while the elastic magnetic field is like a compressed spring. The remnant of the photon's kinetic energy is stored in the form of angular momentum that presses it hard against the nucleon, making it possible for the photon to rest on the nucleon. The photon will rest temporarily if conditions are favourable: if its other forms of energy, especially magnetic energy, and charges harmonize with that particular patch of the nucleon.

But if conditions are hostile, the photon is catapulted away after the collapse of its elastic magnetic field. The catapulting motion transforms its energy back to kinetic energy and accelerates the photon away from the metallic atom.

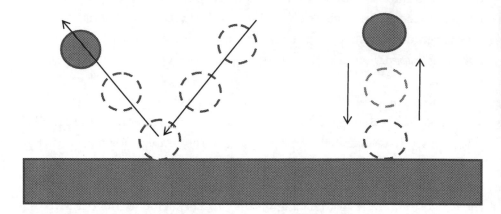

Figure 39: Moving billiard balls being deflected by a piece of wood

Let's say a dynamic photon collides with a stationary photon that is perched on the nucleon. A build-up of an elastic magnetic field between the two photons will take place. How well the magnetic field builds up between them hinges on how well the stationary photon adheres to the nucleon.

If the stationary photon does not adhere strongly to the nucleon, the stationary photon may easily be dislodged and transform to a dynamic photon again. This causes the incident dynamic photon to be deflected randomly. There is no guarantee that the angle of incidence and the angle of reflection will be exactly the same.

If the stationary photon adheres strongly to the nucleon, a smooth build-up of an elastic magnetic field between the dynamic photon and the stationary photon takes place. This ensures that the angle of incidence is exactly the same as the angle of reflection for the dynamic photon.

The interaction between dynamic photons and stationary photons on a nucleon is very complex. Not all dynamic photons are deflected. Such interaction is considered random. We have yet to figure out how the build-up of an elastic magnetic field takes place between the dynamic photon and the stationary photon to ensure the symmetry between the angle of incidence and the angle of reflection.

The magnetic field of a photon is not homogenous. There are patches of magnetic fields on a photon that vary in strength during the build-up of an elastic magnetic field. Such variation enables most incident dynamic photons to deflect such that their angle of incidence is almost identical to their angle of reflection.

The way that light can be reflected by a metallic plate suggests that the metallic nucleus must be saturated with stationary photons. Those stationary photons must adhere strongly to the nucleons so that the reflected light appears as a mirror image.

Lastly, the conundrum of the build-up of an elastic magnetic field between the dynamic photon and the nucleon is yet to be unravelled completely. Does this imply that a photon consists of smaller subcomponents? Is the photon a homogenous unit? The emergence of patches of magnetic fields when the photon interacts with the nucleon is one of its unique characteristics. The quest to better understand this seemingly simple and straightforward phenomenon is still far from resolved.

## Refraction, Dispersion, and the Geometrical Shape of the Unique Nucleus Structure

A needle can float on the surface of still water as if there is a membrane on top of the water. A water molecule is a dipole: the oxygen anion is negatively charged, and individual hydrogen cations are positively charged. Oxygen in a water molecule is attracted to the hydrogen of adjacent water molecules. The hydrogen in a water molecule is attracted to the oxygen of adjacent water molecules. This is what forms the layer of membrane. Dynamic water molecules beneath the membrane move constantly and randomly, propping up the membrane, which acts as a boundary condition for the still water – for example, water in a beaker. Though the needle is much denser than the water, the surface membrane can support the weight of the needle.

If we tap lightly against the wall of the beaker and cause a disturbance of the water membrane, the floating needle immediately sinks to the bottom of the beaker. A physical disturbance of the water membrane cause the static surface water molecules to move randomly, which breaks the membrane apart laterally. The moving water molecules are no longer able to support the weight of the needle.

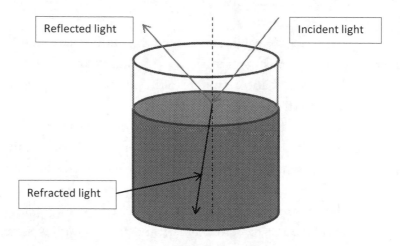

Figure 40: Light refracted by a beaker of water

The surface membrane consists of motionless water molecules that are attracted strongly to one another. The rest of the water molecules in the beaker move freely. Therefore, the density of the water membrane is higher than the density of the rest of the water in the beaker. When the water membrane sinks onto the water in the beaker, those moving water molecules support the weight of the water membrane. Ordinary water molecules are not dense, and they are not moving at very high speeds at room temperature. This explains why water in a beaker forms a meniscus on the top surface of the sagging water membrane.

When incident light is shone on a water membrane at an angle, as shown in Figure 40, the light is reflected by the membrane. The angle of incidence is similar to the angle of reflection. Most of the incident light enters the beaker, where it forms a refraction. The refracted light is close to "normal".

If we move the incident light around on top of the water membrane while maintaining its angle of incidence, its angle of refraction is also maintained. The refracted light is straight in the beaker. We can see the path of the refracted light because dynamic photons of the light interact with dynamic water molecules, causing the light to scatter. We can see the path the refracted light is moving towards.

Incident light striking different portions of the nucleus structure of water molecules on the membrane is reflected differently. When the incident light hits the top portion of the unique nucleus structure, it is reflected such that the angle of incident light is similar to that of the reflected light.

The incident light that hits the lower portion of the unique nucleus structure is deflected to form refracted light. The angle between the "normal" and the refracted light is much smaller than the one between the angle of incidence and the "normal".

The refracted light is straight, which proves that the dynamic photons are not interacting a lot with the dynamic water molecules. The dynamic water molecules are not as saturated as the ones in the membrane. Only a sizeable number of dynamic photons in the refracted light scattered by dynamic water molecules enable us to see the path the refracted light is travelling on.

The incident light that hits the middle portion of the unique nucleus structure enables dynamic photons to scatter and form a bright spot.

How incident light is refracted by the water membrane shows that light is dynamic photons per volume per time, and photons are particles that possess mass, and dynamic photons have momentum. The relationship between the angle of the refracted light and the angle of the incident light clearly proves that the geometrical shape of the unique nucleus structure of water molecules, especially oxygen, is not a perfect spheroid, since the angle of incidence is much larger than the angle of refraction.

When we maintain the incident light angle to the water membrane, we realize that the angle of refraction likewise remains the same. The refracted light has a straight path too. In addition, the angle of incidence and the angle of reflection remain the same. The reflected light and refracted light branch out at the end of the incident light, which strongly proves the continuous flow of dynamic photons. A water membrane has a very organized, compact molecular arrangement. Therefore the relationship between the angle of incidence and the angle of refraction remains the same anywhere on the water membrane.

Refracted light in the water remains on a straight path because the free path among water molecules is rather long. Therefore the dynamic photons of refracted light can travel with minimal interactions with water molecules.

When dynamic photons collide with the nuclei of dynamic water molecules, dynamic photons of refracted light are scattered. This enables us to see the straight path the refracted light is travelling.

As with the beaker of water, most of the incident light shining on the top surface of a block of glass is refracted. Some light is reflected with an angle of incidence similar to that of the angle of reflection. The refracted light is closer to "normal" when it passes through a block of glass. If the angle of the incidence of the light is maintained, and the light is allowed to move anywhere on the top surface of the glass, as shown in Figure 41, then the angles of the normal and refracted light remain the same.

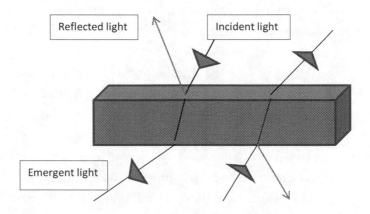

Figure 41: Light refracted by a block of glass

Light entering the block of glass at any point on its upper surface is refracted in exactly the same manner. Light passing through the block of glass is straight, as if there is minimal interaction between the glass molecules and the passing. Scattered light is much brighter in glass than it is in water, which indicates that there are more dynamic photons of refracted light that have interacted with glass molecules. This implies that the molecular arrangement of glass is more compact than dynamic water molecules. It also implies that the boundary condition of a block of glass is somewhat different from that of a beaker of water.

The opposite surface of a block of glass has a similar boundary condition to the top surface of the glass. Light passing through the glass is refracted far away from "normal" into the air after it has exited from the glass, as shown in Figure 41.

The boundary condition of a block of glass is the same on all its surfaces. The angle of incidence for a light shining towards the bottom surface of the block is refracted closer to the normal, and emergent light is refracted far away from "normal" as it exits from the top portion of the glass. Incident light shining towards the top portion of the block passes through. Its loci is exactly the mirror image of the loci of the incident light shining towards the bottom of the glass, as shown in Figure 41.

Incident light is refracted when passing through a block of glass. How close the refracted light is towards "normal" hinges on the angle of incidence as well as the density of the block of glass. There is a direct association between the unique nucleus structures of the glass molecules and their density. The denser the unique nucleus structure, the denser the glass, assuming that the saturation of molecules per volume is the same, which may not necessarily be true. The way incident light is refracted depends on the geometrical shape and size of the unique nucleus structure of the glass molecules.

We can see the path of refracted light passing through the glass because the refracted light is scattered when the light hits the nuclei of glass molecules standing in its path. The refracted light still travels in a straight path through the glass, like the refracted light in a beaker of water. The refracted light interacts with glass molecules at a minimal level, since the density of molecules within the block is slightly lower as compared to the density on the boundary surface. In addition, the saturation of the molecular structure of the block of glass is slightly different from the one at the boundary. Therefore, the glass molecules allow light to pass through the glass in a straight path. The uncanny capability of glass to transmit light in straight path is still unresolved.

We know with certainty that the speed of light is much slower when light is transmitted through a block of glass. Therefore the molecular arrangement within the block is different from the one at the boundary, which could help to explain why light is transmitted in a straight path within the glass. Some light does scatter when it collides with the glass molecules. The path of light passing through the glass glows, which can be seen from all different directions.

Like a prism, a block of glass can scatter refracted light into a spectrum when incident light strikes the surface at a specific angle of incidence. Dynamic photons in the light collide with the unique nucleus structure of the glass molecules on the boundary condition and disperse into a spectrum, as shown in Figure 42.

It is believed that there are two invisible lights at either end of the spectrum. They are the ultraviolet and the infrared. Quantum scientists believe red light has the longest wavelength and blue light has the shortest wavelength. Ultraviolet has an even shorter wavelength than blue light. Infrared has an even longer wavelength than red light.

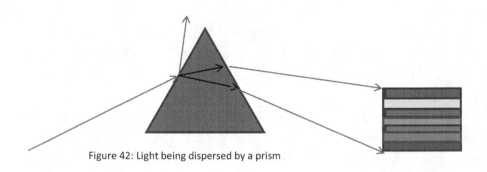

Figure 42: Light being dispersed by a prism

The scientist Sir Frederick William Herschel, in the eighteenth century, measured the temperature of each wavelength of light in the spectrum. He realized that the temperature rose sharply when he placed the thermometer "beyond" red light, a point he called *infrared*. This experimental proof implies that the infrared possesses a higher saturation of dynamic photons per volume per time than red light. Why?

The nature of a dispersed light depends upon the molecular structure, size, and geometrical shape of the unique nucleus structure of the material of the prism, as shown in Figure 43.

We can only say that this is how a glass molecule deflects dynamic photons of the incident light. It disperses the photons before transforming the light into a spectrum. Both ends of the spectrum are sandwiched with invisible light: the ultraviolet and infrared. Ultraviolet light has a higher saturation of dynamic photons per volume per time than blue light. Infrared light has a higher saturation of dynamic photons per volume per time than red light. So it is wrong to believe that infrared light is much weaker than red light.

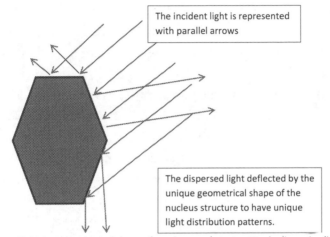

The incident light is represented with parallel arrows

The dispersed light deflected by the unique geometrical shape of the nucleus structure to have unique light distribution patterns.

Figure 43: Unique geometrical shape of a unique nucleus structure in dispersing light

If a prism is made of a material other than glass, like diamond, we expect that light passing through the prism will be dispersed into a different kind of spectrum than that of a glass prism.

## Red Shift and Blue Shift

A receding star appears slightly reddish, while an approaching star appears slightly bluish. Quantum scientists believe that these colours are due to the Doppler effect. A receding star's light is stretched so that it appears slightly reddish. An approaching star's light is compressed, which makes it appear slightly bluish. Scientists make the analogy to the siren of a patrol car, which sounds different when the sound approaches the observer versus when the sound is receding. Are the scientists correct?

Let's say a patrol car with its siren on is stationary, at a point five hundred metres away from an observer at the beginning. The car approaches the observer in increments of fifty metres, stopping for one minute at each interval. Every time the patrol car moves closer to the observer, the pitch of its siren seems to become louder and sharper. When the patrol car stops momentarily, the pitch of the siren stabilizes.

This observation clearly proves that the closer the patrol car gets to the observer, the less work the sound waves do to reach the observer. The sound waves retain more energy as the distance gets shorter. Thus, the pitch of the siren becomes sharper and louder as the distance decreases. If the patrol car moves in this manner towards an observer, there is no Doppler effect on the pitch of the siren.

In another experiment, the patrol car starts from the same spot as the observer. This time, the patrol car recedes fifty meters each time, stopping for one minute at each increment. As the patrol car moves farther from the observer, the pitch of the siren seems to get flatter and weaker. As the distance between the patrol car and the observer increases, the sound waves need to do more work to reach the observer. The sound waves retain less energy and seem weaker. Again, there is no Doppler effect on the pitch of the siren.

So how do we explain the change in colour of stars, by which approaching stars turn bluish and receding ones turn reddish?

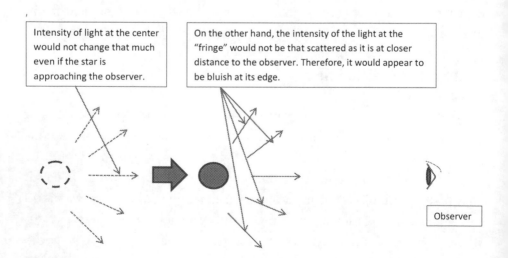

Figure 44: Blue shift

The centre of an approaching star doesn't demonstrate an obvious change in colour, but the edges appear bluish. This is because the intensity of light at the centre does not change much, but the light at the edge is not as scattered as before to the observer. Therefore it appears bluish, as shown in Figure 44. Figure 44 is a little exaggerated, but the diagram is succinct.

Light from an approaching star does less work to reach the observer as it travels a shorter distance, conserving energy. Therefore the light from an approaching star in general should appear bluish. But in reality, only the edge of the approaching star appears bluish. Maybe the energy that has been saved by doing less work is insignificant.

A receding star appears reddish, especially at the edges, as shown in Figure 45. The light at the edges appears reddish because it is more scattered than before as it moves farther away from the observer.

The centre of a receding star appears the same because the light emitted from the centre portion is not scattered. In general, the receding star appears reddish, especially at the edges.

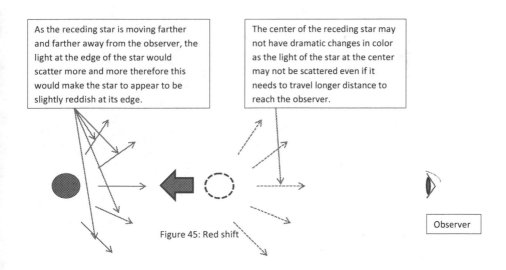

As the receding star is moving farther and farther away from the observer, the light at the edge of the star would scatter more and more therefore this would make the star to appear to be slightly reddish at its edge.

The center of the receding star may not have dramatic changes in color as the light of the star at the center may not be scattered even if it needs to travel longer distance to reach the observer.

Figure 45: Red shift

Observer

Since the light of a receding star does extra work, travelling a longer distance to reach the observer, it uses more energy. The additional distance that the light needs to travel is insignificant and doesn't really affect the colour of the receding star, especially at the centre.

Provided the star is travelling at great speed, receding or approaching, the travel causes its energy to either do more work to travel a longer distance to reach the observer, or do less work to travel a shorter distance. The differences in energy expended are not large, but significant enough to cause visible changes in the colour of the light reaching the observer, especially at the edges of the star.

A receding star does not appear reddish because its light is stretched, nor does an approaching star appear bluish because its light is compressed. Light is dynamic photons per volume per time. Changes in the saturation of dynamic photons per volume per time lead to changes in perceived colour. Each colour has a distinctive range of dynamic photons per volume per time.

# CHAPTER 4

# ELECTRONS

## Intrinsic Spin, Stationary Photons, and Electrons

All scientists believe that there is only one type of electron – in other words, all electrons are identical. So what make free electrons different from stationary electrons? Stationary electrons are confined to an atom, whereas free electrons drift from atom to atom.

Anderson discovered another particle that is identical to an electron in term of its mass in a cloud chamber. But he believed this particle to be positive in charge, and named it *positron*. The word *positron* is derived from *electron* and means a positively charged electron. Anderson's conclusion was based on his observation of the path that the so-called positron traversed, which was opposite that of most electrons. Because of this opposite path, Anderson believed that such a particle should possess the opposite charge. Electrons carry a negative charge, so he assigned positrons a positive charge. Is it true that a positron is a positively charged particle with a mass equal to an electron?

The interaction between the magnetic field of a cloud chamber and the magnetic field of a particle produces a particular outcome. A particle travels in a particular direction in a cloud chamber. It is right to assume that an opposite-charged particle possesses an opposite magnetic configuration. But we don't know with certainty that a positron possesses a positive charge. A similarly charged particle with the opposite intrinsic spin direction may also display the opposite magnetic configuration.

If a positron is repelled by a proton, then we can be certain that a positron possesses a positive charge. Or a positron could be a negatively charged particle like an electron, but with an opposite intrinsic spin to most electrons, giving it an opposite magnetic configuration to that of most electrons.

When a switch is turned on, a light bulb lights up brightly, as shown in Figure 46. Batteries supply free electrons that light up the bulb.

Light consists of dissipating photons with a certain saturation per volume per time. Dynamic photons emitted from the lit bulb should be derived from the free electrons flowing through its filament.

Free electrons must be saturated with *stationary* photons. Why do *dynamic* photons dissipate from free electrons when they flow through the filament of a light bulb?

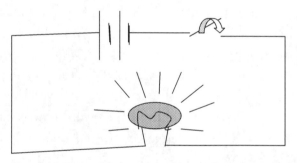

Figure 46: Emitted light is derived from free electrons

Tungsten filament has high resistance that hinders the flow of free electrons. When free electrons experience stagnation in their flow, they transform their stockpile of stationary photons to dynamic photons, which are released as heat and light.

Free electrons are not ordinary electrons. They contain a higher saturation of stationary photons. Therefore, free electrons are more massive and have higher momentum. Consequently, they can drift from atom to atom.

A piece of wire has low resistance because metallic atoms have large electron shells and the overlapping among those electron shells is extensive.

Free electrons can flow through such a wire without losing many of their stationary photons.

Atoms of an insulating material have much higher resistance because the atoms have a much smaller electron shell. This prevents free electrons from flowing through them easily. The free electrons are forced to dissipate most of their stockpile of stationary photons and are captured by the insulating atoms as stationary electrons.

When there is a current flowing through a circuit, as shown in Figure 47, the entire circuit is enveloped in a magnetic field. Different circuit configurations have magnetic fields with different orientations and configurations. Current flow patterns shape the magnetic field. Each free electron is like a tiny magnet on the move. A collection of them flowing in a particular pattern per time shapes a specific configuration of the magnetic field.

A stockpile of stationary photons on a free electron not only gives it higher momentum, but also makes it act like a stronger magnet. A much weaker magnetic field envelops a stationary electron. This implies that stationary photons have much stronger magnetic fields than dynamic photons. In addition, free electrons are saturated with more stationary photons than are stationary electrons. Therefore, an individual stationary photon acts like a tiny magnet. A collection of them superposes to form a stronger magnetic field enveloping a free electron.

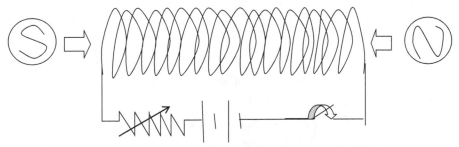

Figure 47: The presence of current gives rise to magnetic field

With a stronger current flow, a circuit is enveloped in a stronger magnetic field. This clearly attests that the magnetic field enveloping the circuit is directly derived from dynamic free electrons. A larger number of free electrons flowing through a circuit ensures it is enveloped in a stronger elastic magnetic field per time.

## Positrons and Electrons

In regard to Anderson's discovery of the positron, it is true that all atoms are enveloped by their respective stationary electrons in the electron shell. This applies to all air molecules in a cloud chamber at high altitude. Sunlight is nothing but a cluster of dissipating dynamic photons per volume per time. At higher altitude, sunlight is more intense.

In Anderson's experiment with a cloud chamber at high altitude, he occasionally observed that an atom of an air molecule was ionized in a way that left behind tracks of cloud that were symmetrical to ones left behind by an electron under similar influence from a fixed magnetic field. However, the tracks ran in a contrary direction. Thus, he claimed the tracks has been demarcated by a positron, which he believed had an equal mass to an electron but an opposite, positive charge. His assumption was that only an opposite charged particle reacts in an opposite manner to a free electron when interacting with a similar fixed external magnetic field in the cloud chamber. Is this assumption true?

It is highly unlikely that a so-called positron is positively charged, since all atoms are enveloped by negatively charged stationary electrons in the electron shell. Protons and neutrons are bound within the nucleus. Ionization of an atom occurs when some electrons dissipate from its electron shell under the influence of strong sunlight. This reassures us that a positron must also possess a negative charge. So why does a positron leave behind an opposite track symmetrical to that of a free electron?

Apparently, the presence of more intense sunlight at higher altitude initiates ionization in air molecules. Stationary electrons modify their capacity to stockpile stationary photons. Stockpiles of stationary photons within the nucleus are also subjected to modification at different altitudes due to changes in the effect of localized gravitational potential energy.

Under the conditions of Anderson's experiment, observation of a positron spurting from an atom in an air molecule is rarer than observation of free electrons. This suggests that electrons outnumber positrons on earth.

We are almost certain that there is only one type of electron and all electrons are identical except in their stockpiles of stationary photons. (We will justify this assumption when we discuss how current is produced.) We conclude that the occurrence of a positron track in a cloud chamber is the product of pure chance. The unique adherence of stationary photons to a positron in a similar interaction between a stationary positron and intense sunlight in a strong, fixed, external magnetic field causes the positron to be enveloped with an opposite magnetic field to that of an ordinary free electron. Such a positron is likely to interact with the fixed external magnetic field to produce a completely opposite outcome: an opposite track that is symmetrical to the one left behind by ordinary electrons.

Some may find this explanation hard to digest. Then again, we still have a shallow understanding of magnetism. All magnets have two poles, north and south. We can break a magnet down into much smaller segments, which themselves are tiny magnets. There are magnetic fluxes surrounding a magnet. We believe magnetic fluxes move out from the north pole and drift towards the south pole. Generally, magnetic fluxes are more saturated at the magnetic poles.

Two or more magnetic fields merge to form an elastic magnetic field. We don't have evidence as to how this merger take place, except for observations of the behaviour of ferrous powder. Ferrous powder poured on a piece of paper above a magnet and then tapped takes on the shape of the newly formed elastic magnetic field. The visible loops in the ferrous powder formation give us no clue about its magnetic polarities.

We are in this predicament because the magnets that we know have two poles. Anything that is different from a dipole configuration bogs us down miserably.

It is possible that a positron may be enveloped in an opposite elastic magnetic field to that of an ordinary free electrons. Thus, a dynamic positron swerves in an opposite track from that of a free electron, as a result of interaction with a fixed external magnetic field. This doesn't imply in any way that a positron is positively charged.

A free electron and a single proton swerve in opposite directions when they interact with a fixed external magnetic field because the free electron is negatively charged, while the proton is positively charged. But a dynamic neutron drifts in a similar direction to the free electron, and we are sure that a neutron is neutral in charge.

This example attests to the magnetic-magnetic interaction between a dynamic particle and a fixed external magnetic field, which in turn determines the particular direction that the particle drifts as a result of the interaction. This reinforces our belief that positrons are actually electrons with an opposite intrinsic spin direction, which gives the positron an opposite magnetic configuration to that of an electron.

Let's say a dynamic photon is also an even tinier magnet with a specific orientation. The red arrow indicates its intrinsic spin direction, and the vertical green arrow indicates its angular momentum, which determines its direction of motion, as shown in Figure 48.

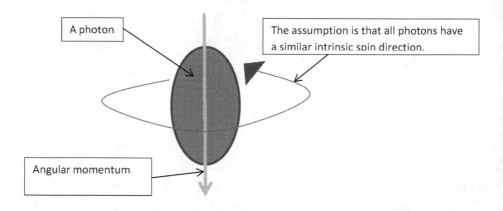

Figure 48: A photon with its angular momentum and intrinsic spin direction

We assume that there is only one type of photon. All photons have a single, unique intrinsic spin direction. In other words, all photons have the same intrinsic spin.

Let's say a photon is positioned with its green arrow pointing towards an electron. It adheres to the electron as shown in Figure 48. Since the magnetic polarities of the electron and the photo are comparable, the stockpile of stationary photons adhering to the electron boosts its magnetic strength.

Since a positron has an opposite magnetic polarity to that of an electron, we expect that the photon will re-orient itself with its green arrow pointing upward and its tail adhering to the positron. This enhances the magnetic strength of the positron. If a sizeable number of stationary photons re-orient themselves when they adhere to a positron, that will greatly boost the opposite magnetic strength of the free positron. Therefore, a free positron that has amassed sufficient stationary photons has a stronger but opposite magnetic configuration to a free electron.

Generally, a bare positron has an opposite magnetic configuration and orientation to that of a bare electron. A free electron has a similar magnetic configuration to a bare electron. The same should apply to positrons too.

In conclusion, the orientation of stationary photons adhering to free electrons should be opposite to that of stationary photons adhering to a positron. Therefore the bond between stationary photons and a positron is much weaker than the bond between stationary photons and an electron.

Because of this weaker bond, a positron will not amass as many stationary photons as an electron does. Generally, a free positron is much less massive than a free electron. Fewer stationary photons perch on a free positron than on a free electron.

Anderson's experiment clearly shows that the detected positrons are saturated with a sizeable number of stationary photons, and the bond between them is strong. This may suggest the possibility of two distinct types of photons, each with a different intrinsic spin. But this has not been proven. So, for the time being, we will stick with the current school of thought that says there is only a single type of photon.

Positrons and electrons are negatively charged particles with different intrinsic spin directions. Therefore, positrons have a different magnetic configuration from electrons. Free positrons interact with a fixed external magnetic field with an opposite outcome than that of free electrons interacting with a similar field.

## Interaction Between Dynamic Photons and an Electron

The velocities of stationary electrons and free electrons are much slower than the velocity of dynamic photons. Therefore, a dynamic photon adhering to a stationary electron conserves its kinetic energy by transforming it to other forms of energy. A dynamic photon that converts to a stationary photon has not lost energy, but has converted it instead.

We believe a stationary photon resting on a stationary electron hastens the electron's intrinsic spin rate, fostering the production of a stronger magnetic field. Subsequently, this stronger magnetic field merges with the magnetic field of the stationary electron to form an elastic magnetic field. The elastic magnetic field acts like a compressed spring locked between the stationary photon and the stationary electron.

Whether a dynamic photon adheres to a stationary electron is a matter of chance, as illustrated in Figure 49.

A stationary electron drifts in a much larger ellipse from left-hand side to the centre of the page, as indicated by an arrow that happens to intercept a dynamic photon.

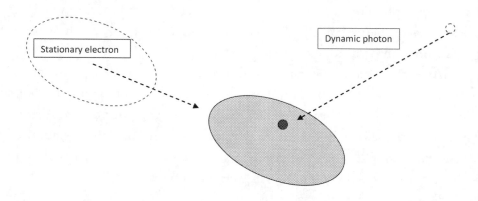

Figure 49: A bare stationary electron intercepts a dynamic photon

Let's assume that the stationary electron is bald. Any photon can adhere to it readily at any point. This condition increases the chance that a dynamic photon will adhere to it, provided the portion of the stationary electron where the photon adheres has an opposite charge to that of the photon,

We are assuming that a photon is a charged particle. Free electrons have a stronger affinity for the anode of a cathode tube than do ordinary electrons, which indicates that a photon is a negatively charged particle. Free electrons that are more saturated with stationary photons behave as if they are more negatively charged; therefore the photon must be negatively charged. This suggests that the electron is not a homogeneous, single piece. It also contains a positive quark that stationary photons can adhere to. In addition, we believe that only a charged particle can transform its kinetic energy to magnetic energy by increasing its intrinsic spin. This fosters a stronger magnetic field that envelops the photon.

The superposition of the magnetic field of an electron and that of a stationary photon forms a stronger elastic magnetic field. Such an elastic magnetic field collapses before the emergence of an acting force against the stationary photon. The action of that force triggers the photon to transform back to a dynamic photon. Its kinetic energy returns to status quo.

In another example, let's say there is a stationary photon already resting on a stationary electron, as shown in Figure 50. Another dynamic photon tries to adhere to the same spot. The dynamic photon will collide with the stationary photon when the dynamic photon intercepts the stationary electron. The collision is likely to dislodge the stationary photon from the stationary electron, transforming it to a dynamic photon. As a result, the intercepting dynamic photon bounces away, and two dynamic photons emerge from the collision.

If the stationary photon adheres very strongly to the stationary electron, then the dynamic photon can bounce away without dislodging the stationary photon.

Just before a stationary photon transforms to a dynamic photon, it converts its compressed elastic magnetic field to an acting force. The force acts against the photon and enables it to accelerate, strengthening its angular momentum. More angular momentum acting on the dynamic photon increase its velocity. This is how a stationary photon transforms its magnetic energy back to kinetic energy.

We reckon that less than 20 per cent of the total energy of a dynamic photon is recycled. The interchange between magnetic energy and angular momentum accelerates the photon and enables it to travel intergalactically. Therefore, a dynamic photon is not enveloped by a strong magnetic field the way a stationary photon is.

Stationary photons on stationary electrons are usually enveloped by a strong magnetic field after they have transformed most of its kinetic energy to magnetic energy, in the form of an elastic magnetic field. The elastic magnetic field represented a merging of the magnetic field of the stationary photon and that of the stationary electron. Free electrons are enveloped in strong magnetic fields because they are saturated with stationary photons.

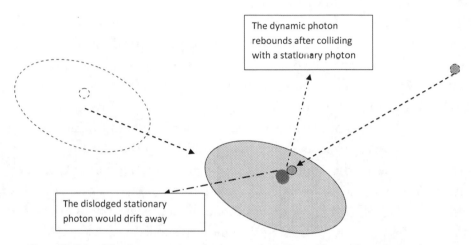

Figure 50: The collision between a dynamic photon and a stationary photon of a stationary electron

Let's say there is a cluster of stationary photons already resting on a drifting stationary electron, as shown in Figure 51. In addition, there is a dynamic photon intercepting the stationary electron at a vacant spot on the stationary electron.

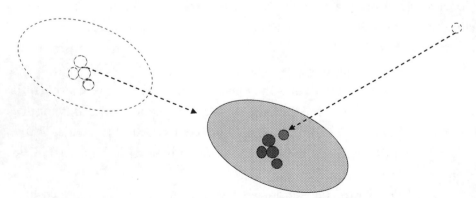

Figure 51: A dynamic photon lands on a vacant spot of a stationary electron

Once the dynamic photon intercepts the stationary electron at its vacant spot, the dynamic photon is likely to be transformed to a stationary photon, provided the newly forged elastic magnetic field around the cluster of stationary photons harmonizes with the potential stationary photon. The vacant spot must also have an opposite charge to that of the photon. Compatibility makes way for a soft, smooth landing for the photon. It transforms kinetic energy to hasten its intrinsic spin, creating a stronger magnetic field that merges harmoniously with the elastic magnetic field of the cluster. This enables the dynamic photon to complete its transformation into a stationary photon and adhere strongly to the stationary electron.

To make things a little more complicated, all stationary electrons possess their own intrinsic spin, and they may be spiralling as they thrust forward. So stationary electrons do not drift quite as illustrated in preceding examples.

We believe that all electrons possess intrinsic spin. All bare electrons are enveloped in a magnetic field. But we are uncertain how they are actually spinning. To make thing easy, we assume that all electrons are oblong spheroids in shape, and that they spin clockwise on the longitudinal axis of their angular momentum, as shown in Figure 52.

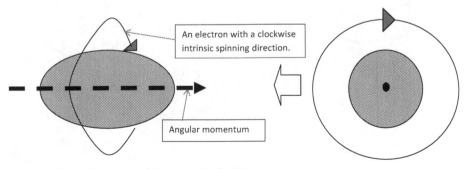

Figure 52: Electron and its intrinsic spin direction

Since all stationary electrons rotate clockwise on their own longitudinal axis of angular momentum, the interactions between stationary electrons and dynamic photons take place randomly. Interactions not only depend on the movement of the stationary electrons but also on the availability of dynamic photons that happen to be intercepted by the electrons. In addition, stationary electrons are enveloped by their own magnetic fields, which derive from their intrinsic spin.

Positrons have a negative charge similar to that of electrons, except their intrinsic spin is in the opposite direction: counterclockwise, as shown in Figure 53. Therefore, positrons are enveloped by a magnetic field of the opposite orientation and configuration to that of electrons. Positrons can easily be wrongly identified as positively charged, because they drift in the opposite direction of electrons after they have interacted with a fixed external magnetic field in a cloud chamber.

A cathode tube with a fixed external magnetic field is a perfect apparatus for proving that all electrons are identical to one another. Maybe we should redo the experiment to detect the presence of positrons at the same site where Anderson first detected them. Maybe at that particular altitude (e.g., a specific level of localized universal gravitation potential energy), positrons are effective in stockpiling stationary photons, which increases the chance of detecting their presence.

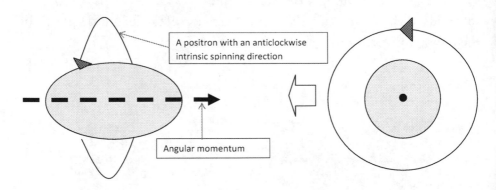

Figure 53: Positron and its intrinsic spin direction

Maybe there are only a small number of positrons present on earth, making them harder to detect even armed with a sophisticated cloud chamber. Maybe positrons are not effective in stockpiling stationary photons at sea level, making it even more difficult for them to undergo an intense exchange of photons. A positron has an opposite magnetic configuration to that of an electron, making it harder for a positron to retain stationary photons (if there is only one type of photon).

All free electrons emitted from an electron gun and interacting with a fixed external magnetic field end up in a similar spot on the cathode-ray tube's screen, which clearly attests that there is only one type of electron. Unfortunately, positrons always fail to access electron guns. An electron gun is not a good device for detecting the presence of positrons.

We know with certainty that all photons have intrinsic spin because stationary photons are enveloped in stronger magnetic fields than dynamic photons. Since photons can interact with both positrons and electrons equally well, it's possible there are two types of photons, which in turn may possess two distinctive types of intrinsic spin. More research needs to be carried out before we can validate this hypothesis.

## Electron Shell and Edgorbtoslengths

A cluster of atoms at rest on top of one another will not experience a collapse of their electron shells. They can resist the pressure of their own weights. This is because mosaic of magnetic field of the nucleus props the electron shell and prevents it from caving into the nucleus. The opposing forces between the mosaic of magnetic fields of the nucleus and that of stationary electrons are stronger than the mutual attractive electrostatic forces between stationary electrons and protons within the nucleus of an atom. Therefore the electron shell is sturdy.

Stationary electrons drift at great speeds and are kept within the electron shell by mutual attraction between stationary electrons and protons within the nucleus. The electron shell would be flimsy if not for the mosaic of magnetic field of the nucleus that props the electron shell. Otherwise any acting force on the electron shell would drive the stationary electrons towards the nucleus.

A stationary electron drifts more slowly than a dynamic photon. The balance of mutually attractive forces between the proton and the stationary electron and the opposing forces between the magnetic field of the stationary electron and the mosaic of magnetic field of the nucleus keeps the stationary electron within the electron shell. A stationary electrons that has gained sufficient energy by stockpiling stationary photons to drift away from the nucleus is called a *free electron*.

Some electrons can drift away from an atom to adjacent atoms without stockpiling stationary photons. This occurs when the atom is under external pressure. Such electrons are called *drifting electrons*. Some materials can produce sizeable numbers of drifting electrons, especially when their electron shells overlap extensively. Drifting electrons move freely from atom to atom, much like free electrons, but they don't possess significant momentum and have a stockpile of stationary photons more or less similar to stationary electrons. Therefore, they don't possess enough power to do work.

Basically, stationary electrons surf on top of such mosaic of magnetic field of the nucleus. The mosaic of magnetic field and that of stationary electrons form an elastic magnetic field per time. Changes in such an elastic magnetic field are likely to affect the adherence of stationary photons to those stationary electrons.

Each atom has only one electron shell. It demarcates the motion loci of all stationary electrons within the atom. The shape and size of the electron shell is determined by the total stationary photon stockpile within the nucleus of the atom and the uniqueness of its nucleus structure.

Except for the single-proton hydrogen atom, the nuclei of elements consist of both protons and neutrons. These nucleons arrange themselves in a unique nucleus structure to maintain cohesiveness and stability. Each element has a specific number of protons and neutrons. These nucleons arrange themselves in a unique nucleus structure, in which protons do not sit close to one another. The presence of neutrons diffuses the tension among protons. Electrons provide interlocking attraction to those protons within the unique nucleus structure.

As the elements get denser, the increment of the number of neutrons is outnumbered by the increment of the number of protons in the elements' nuclei. The increasing number of protons requires more neutrons to diffuse tension among the protons.

All nucleons possess intrinsic spin and are enveloped by their own magnetic field. Protons and neutrons have different magnetic orientations and configurations. Therefore, we believe that a unique nucleus structure forms a mosaic of elastic magnetic field that consists of the superposed magnetic fields of all protons and neutrons within the unique nucleus structure.

Protons and neutrons also have their own stockpiles of stationary photons. Their stockpiles not only affect their mosaic of magnetic field but also help determine the size of the electron shell through flexure of the unique nucleus structure. For instance, if there is a higher stockpile of stationary photons within the nucleus, this makes the mosaic of magnetic field of the nucleus more refined. Imagine the photon is a tinier magnet. The unique nucleus structure rebalances itself through the increased angular momentum of the nucleon. Subsequently, a new configuration of mosaic of magnetic field of the nucleus reshapes and resizes the electron shell.

We have to examine an atom based on the gravitational attraction between its nucleus and its stationary electrons. We must figure out the possible mosaic magnetic field of the nucleus propping the electron shell, the extent of electrostatic attractive forces between the stationary electrons and the protons in the nucleus, and the repulsive forces among protons within the nucleus. Changes in the stockpile of stationary photons on stationary electrons and nucleons not only modify their angular momentum but also upset the balance of gravitational, electrostatic, and magnetic forces within the atom.

Changes to the stockpile of stationary photons on the stationary electrons and the nucleons cause modification of their angular momentum immediately. Denser nuclei exert a stronger gravitational attraction on their stationary electrons. If the stationary electrons are also denser, the mutual gravitational attraction between the nucleus and those stationary electrons gets even stronger.

At the same time, dense stationary electrons have more momentum, allowing them to drift farther away from the nucleus. The mosaic of magnetic field of the nucleus gets even more refined with the superposition of the magnetic field of individual stationary photons on the magnetic field of all the bare nucleons.

The presence of stationary photons on protons tempers their electrostatic strength. Stationary electrons that have larger stockpiles of stationary photons strengthen their negative electrostatic strength. A sizeable change in the stockpile of stationary photons within an atom has severe effects on the atom in term of its physical and chemical characteristics.

Stationary electrons that possess greater stockpiles of stationary photons are able to drift much farther away from the nucleus due to their higher momentum. If their momentum gets very strong, they may be able to drift from atom to atom, upon which they are reclassified as free electrons. How much a stationary electron changes its stockpile of stationary photons depends upon the condition on the mosaic of magnetic field of the unique nucleus structure.

Free electrons are enveloped in a strong magnetic field derived mainly from its stationary photons. Free electrons are more massive than stationary electrons and have higher momentums too. Stationary photons also boost the angular momentum of free electrons, partially because a dynamic photon transforms some of its kinetic energy to angular momentum in the process of becoming a stationary photon. A big portion of its kinetic energy is conserved as magnetic energy through the hastening of its intrinsic spin. Most of the magnetic energy that envelops a free electron is derived from · the stationary photons that adhere to the free electron. Its own magnetic energy is partially transformed to angular momentum, which increases.

Generally, stationary electrons and nucleons increase their stockpiles of stationary photons when the temperature rises. This allows a more intense exchange of photons among nuclei. Subsequently, the nuclei stagger apart, expanding their volume. A less intense exchange of photons allows nuclei to come closer to one another, which is why there is shrinkage when the temperature drops.

Table 11 - Density of water, at standard sea-level atmospheric pressure

| Temperature | Density(g/cm$^3$) | Temperature | Density(g/cm$^3$) | Temperature | Density(g/cm$^3$) |
|---|---|---|---|---|---|
| 0°C | 0.99987 | 10°C | 0.99978 | 26.7°C | 0.99669 |
| 4°C | 1.00000 | 15.6°C | 0.99907 | 32.2°C | 0.99510 |
| 4.4°C | 0.99999 | 21°C | 0.99802 | 37.8°C | 0.99318 |

Water is densest at 4° C, as shown in Table 11. It is easy to explain why there is a reduction in water density when the temperature rises. The exchange of dynamic photons intensifies because of the increased availability of dynamic photons from a heat source. The nucleus structure of water molecules experience a net gain in their stockpile of stationary photons.

This directly increases their momentums and speeds. Faster-moving water molecules have longer free paths. As the water molecules gain momentum, they scatter more to minimize collisions among themselves. This creates a longer free path. Thus, the density of water decreases as the temperature increases above 4° C.

Continuous reduction in the stockpile of stationary photons within a water molecule occurs as a result of reduction in temperature until 4° C is reached. The free paths among water molecules shorten because of their concomitant reduction in velocity. Water density increases to its maximum at 4° C. Flexed water molecules at this temperature experience a reduction in effective collisions. They fail to disperse effectively, which causes water to reach its highest density at 4° C.

As the temperature continues to drop below 4° C, water molecules continue to lose their stockpile of stationary photons. This causes them to have weaker impulses during their collisions, since their momentums have weakened further. This may reduce their capability to disperse during collisions, resulting in even shorter free paths.

The density of water below 4° C is actually slightly less than its density at 4° C. The only plausible rationale for this effect is that water molecules regain their effectiveness in dispersing during their collisions due to the flexure of the water molecule. Water molecules may stagger farther apart below 4° C.

It is also easy to explain why ice has slightly lower density than the water at 4° C. Flexing of the nucleus structure of the oxygen atom of the water molecule leads to flexing of the water molecule as a whole. This causes hydrogen atoms in the water molecule to transfer stationary electrons to the oxygen atoms just before water turns to ice. The hydrogen atoms become more positive and the oxygen atoms become more negative. The ionized water molecules bond together strongly to form a solid under the influence of the van der Waals forces among them. This allows the formation of a crystalline structure before the water solidifies to ice at 0° C. The density of ice is slightly less than that of water because there are plenty of voids in the crystalline structure.

Water expands as it turns to ice, which clearly shows that the water molecules stagger closer to one another as they arrange in a crystalline structure. This supports the notion that there are plenty of voids among icy water molecules.

Though ice is slightly lighter than water, ice floats because water molecules that have temperatures above $0°C$ tend to possess higher kinetic energy, since they have higher stockpiles of stationary photons. The impact of water molecules against the ice allows the ice to float. Ice forms on the surface first.

Stationary electrons are frequently attracted to the portions of the unique nucleus structure where there are slightly higher saturations of protons. That space is where regular motion loci of stationary electrons are located. We call this *edgorbtoslength*, which is an acronym of *edg*e of *orb*ital *to* electron *s*hell *length*.

The edgorbtoslengths of an atom are where sharing of electrons is likely to take place. Overlapping of edgorbtoslengths to edgorbtoslengths among atoms enables bonds to be forged. Another name for edgorbtoslength is *bondbital*.

The measure of an edgorbtoslength is used to gauge the strength of bonding between atoms. A longer and wider edgorbtoslength is more porous, allowing stationary electrons from both atoms to swipe through the shared edgorbtoslength with minimal interaction.

Let's say there are three different types of materials, namely metallic, metalloid, and insulator. There are ten stationary electrons passing through each of their shared edgorbtoslengths per time. If the edgorbtoslength of the insulator atom is rather short, then the overlap of the short lengths of shared edgorbtoslengths between two insulator atoms is brittle, because twenty stationary electrons are streaming through a more congested "airspace". This intensifies the interactions among those stationary electrons and is likely to hamper the smooth sharing of electrons among those insulator atoms.

An insulator has too many stationary electrons streaming through a smaller airspace per time. The nuclei of its atoms exert a strong attraction on the stationary electrons within their shorter shared edgorbtoslengths. With too many stationary electrons sharing a small airspace, the smooth sharing of stationary electrons is easily disrupted, making such material weak and brittle.

Metallic atoms have a much longer edgorbtoslength. Overlap among their longer shared edgorbtoslengths diffuses the tension among those stationary electrons. They swipe through the large space without much disruption, which allows the forging of stronger bonds among metallic atoms.

Generally, metallic atoms are denser and have a larger number of protons within their nuclei. These protons exert a strong attraction on the large number of shared stationary electrons within their shared edgorbtoslengths. In addition, metallic atoms tend to have a larger electron shell and longer edgorbtoslength. Metallic atoms can tolerate more strain among themselves by allowing more extensive narrowing and elongating on their longer shared edgorbtoslengths when subjected to tensile stress. The longer shared edgorbtoslengths among metallic atoms allows extensive shearing while still maintaining strong bonds. Therefore, metallic materials are very ductile.

The nature of a unique nucleus structure of an atom determines the shape and size of its electron shell as well as the number of its edgorbtoslengths. Chemical study of an element and its compounds is to ascertain the number of its fronts of edgorbtoslengths, and the number of edgorbtoslengths at each front before and after chemical reaction. It is also important to ascertain the lengths of edgorbtoslengths at each front before and after a chemical reaction.

An element's or compound's edgorbtoslengths at each front and length of edgorbtoslengths differ at different temperatures. For instance, most metallic oxides eventually reduce to pure metallic atoms by releasing oxygen gas into the air if they have been heated to a high enough temperature. This clearly attests that heat causes disruption within the shared edgorbtoslengths while intensifying photon exchange. Heat also hinders hassle-free sharing of electrons, further weakening the bonds among atoms before ripping them apart.

The number of shared edgorbtoslengths at each front and their lengths will be different when an element or compound shares bonds with different types of elements and compounds. The bond strength of different compounds containing atom A differ depending on the other atom it is sharing electrons with. The length of shared edgorbtoslengths of atom A may remain the same at a particular temperature, but the atom that shares electrons with atom A may be different in terms of the length of its edgorbtoslengths. (There is no strict rule regarding this. We have to rely on experimental outcomes to ascertain the actual way electrons are shared. Bear in mind that the unique nucleus structure before and after a chemical reaction is likely to be different as a result of changes in the stockpile of stationary photons within the nucleus)

New ways of sharing electrons among atoms leads to modification of their unique nucleus structure. Heat may be absorbed or dissipated from nuclei that participate in a chemical reaction. It depends on the nature of the electron sharing in their newly shared edgorbtoslengths, which leads to modification of their unique nucleus structure. The unique nucleus structure is affected in the orientation and configuration of its orbital. The newly shared orbital moulds the newly flexed unique nucleus structure. In short, the unique nucleus structure and bondbitals mutually affect one another.

The value of edgorbtoslengths of an atom at a particular front is stated as a real number to signify the actual number of stationary electrons that are shared between atoms. Every shared electron in quantum theory actually signifies ten stationary electrons that have been shared between atoms. This is because the absolute electrostatic charge of an electron is much weaker than that of a proton, since there are so many different types of super acids with different causticities. We believe that the ratio of absolute charge between a proton and an electron is 1:10. It may be appropriate to put quantum theory's fallacy in perspective by stating the actual number of shared stationary electrons that forge bonding between two atoms.

For instance, 1.2 edgorbtoslengths doesn't really imply that there is a complete edgorbtoslength and another 0.2 fraction of an edgorbtoslength. It signifies that there are twelve stationary electrons that have been shared with a particular atom in that front. The numerical value of an edgorbtoslength is almost always proportional to the strength of the bonding between atoms. The more shared stationary electrons within a shared edgorbtoslength, the stronger the attraction, and thus the bonding, with protons within the nuclei. But that is not always true if the edgorbtoslength is lengthy enough to prevent the shared stationary electrons from colliding frequently among themselves, which ensures cohesiveness in the bonding.

A bonding strength of 2.0 magnitude edgorbtoslengths is much stronger than one of 1.0 magnitude. The bonding strength of twenty shared stationary electrons between two atoms should be stronger than the bonding strength of ten stationary electrons. But it is not true that the bonding of twenty shared stationary electrons is twice as strong as the bonding of ten shared stationary electrons, as bonding strength is also affected by the length of edgorbtoslengths.

Theoretically, the length of an edgorbtoslength shared between two atoms is gauged by applying a tensile strength on those atoms to see how they endure. How long does the shared edgorbtoslength last until it breaks apart? Once we know how enduring the bond is under shearing, then we can assess the length of shared edgorbtoslength.

Generally, it is true that the magnitude of an edgorbtoslength is correlated with its width and length. A 2.0 magnitude edgorbtoslength is much wider and longer than a 1.0 magnitude edgorbtoslength. A wider and longer edgorbtoslength allows more stationary electrons to swipe through the shared edgorbtoslength, minimizing the number of collisions while maintaining stronger bonds among atoms. Generally, a higher magnitude of edgorbtoslength signifies stronger bonding between atoms.

A chemical compound changes its chemical characteristics after a chemical reaction, whether an endothermic or exothermic reaction. This causes dramatic changes to its stockpile of stationary photons within the nucleus. This either weakens or strengthens the angular momentum of its nucleons. As a result, the unique nucleus structure flexes differently before and after a chemical reaction. This leads to changes in its edgorbtoslengths. New ways of sharing electrons can either weaken or strengthen the bonds among atoms of a newly formed compound, depending on the number of shared stationary electrons within the shared edgorbtoslengths as well as the length of shared edgorbtoslengths that overlap. Therefore, the chemical characteristics of a chemical compound are different before and after the chemical reaction. All in all, the magnitude of edgorbtoslengths or the bonding strength of atoms is different before and after a chemical reaction.

For example, before a chemical reaction, the edgorbtoslengths of atom A and atom B are 1 and 1.2 respectively. After the reaction, AB compound has values of 0.8 and 1.6 respectively. The value of edgorbtoslengths of atom A has changed from 1 to 0.8, a reduction of 0.2 as a result of sharing stationary electrons with atom B. Atom B's edgorbtoslengths value has changed from 1.2 to 1.6, an increment of 0.4. The edgorbtoslengths of atom B has undergone modification as a result of its new flexing unique nucleus structure after the chemical reaction. That is likely to cause subsequent modification to its electron shell and channel additional stationary electrons into its shared edgorbtoslengths. Therefore, the bonding strength of the shared edgorbtoslength of atom B of compound AB undergoes further strengthening by sharing more stationary electrons with atom A.

Atom A and atom B have different bonding strengths even when they are in the form of compound AB. This is because the bonding strength of atom A in compound AB is the attraction between protons of the nucleus of atom A and shared electrons in compound AB. The bonding strength of atom B in compound AB is the attraction of protons of the nucleus of atom B and shared electrons in compound AB. Therefore the bonding strengths of atom A and atom B in compound AB are different.

To break the bonding of compound AB proves the proverb "a chain is never stronger than its weakest link". An intense exchange of heat that is strong enough to break the bonding of 0.8 edgorbtoslengths of atom A also breaks the bonding between atom A and B of the compound AB, even without physically breaking the 1.6 edgorbtoslengths of atom B.

The discovery of the different causticness of various forms of super acids strongly suggests that the absolute charge of an electron and a proton are not the same. Quantum theory assumes a hydrogen atom is a single-proton nucleus circled by a single electron. A hydrogen cation is a single proton that is a common representative of acidic characteristics. Under this assumption, all hydrogen cations of various types of mineral acids and super acids should be similar to one another in term of their causticity, since all of them contain a similar single proton. In reality, this is not true.

Different acids with similar molarity possess different levels of causticness, even when ionization of hydrogen cations is not an issue. This clearly attests that so-called hydrogen cations of different types of acids are actually different from one another. This is possible if and only if the absolute charge of an electron and a proton are not the same.

As a matter of fact, the absolute charge of an electron turns out to be much, much weaker than that of a proton. The edgorbtoslength of the hydrogen cation of a super acid is much narrower and shorter than that of a nitrate acid. Therefore the hydrogen cation of a super acid has stronger suction to steal stationary electrons from somewhere to diffuse its tension from a severe lack of stationary electrons. Thus, super acid is more powerful and reactive than ordinary acids, like hydrochloride acid, and can coerce chemical reactions to take place with less reactive molecules, such as organic compounds.

Generally, a hydrogen cation of a super acid has fewer stationary electrons within its electron shell than that of hydrochloride acid. Even a hydrogen cation has an edgorbtoslength, which attests that its nucleus of a single proton is not completely spheroid or symmetrical. Its positively charged quark is somewhat lopsided to the side that faces its edgorbtoslength.

Acids have different causticities. Hydrochloride acid is stronger than sulfuric acid, but nitrate acid is the strongest of the three. Chemists disagree with this claim. They all believe that hydrochloride acid is actually stronger than nitric acid.

We know that concentrated hydrochloride acid or concentrated nitric acid alone do not react with gold, but the mixture of concentrated hydrochloride and nitric acid does react with gold. The chemical reactions are stated below:

$$Au(s) + 3NO_3^-(aq) + 6H^+(aq) \leftrightarrow Au_3^+(aq) + 3NO_2(g) + 3H_2O(l)$$
$$Au_3^+(aq) + 4Cl^-(aq) \leftrightarrow AuCl_4^-(aq)$$

We strongly believe that a hydrogen cation of nitric acid is more caustic than that of hydrochloride acid and can initiate a chemical reaction with gold. There are fewer stationary electrons within the hydrogen cation of nitric acid as compared to that of hydrochloride acid. The more caustic hydrogen cation of nitric acid can "chew" on gold atoms superficially by stealing stationary electrons from them. The reaction starts but is not sustainable because the gold atom remains mostly intact in its crystalline structure even after it has lost several stationary electrons.

Chloride anions in an aqua regia solution forge bonds with a potential gold cation. This slowly and gradually pries the gold atom out from its metallic crystal structure to form $AuCl_4^-$ and expose the gold atom more, facilitating further reactions with the hydrogen cations of nitric acid.

Be and Mg are alkaline elements that turn into cations with an oxidation number of +2 when they bind with an oxygen atom, as shown in the equations below.

$$Be + ½ O_2 \rightarrow BeO \quad \Delta H = -608357 \text{ J/mol}$$
$$Mg + ½ O_2 \rightarrow MgO \quad \Delta H = -600751 \text{ J/mol}$$

When a Be atom reacts with an oxygen atom, they dissipate more dynamic photons (i.e., $\Delta H = -608357$ J/mol) than when an Mg atom reacts with an oxygen atom ($\Delta H = -600751$ J/mol). If we assume the oxygen atom will dissipate a similar number of dynamic photons to the surroundings in both reactions, then it is obvious that the Be atom dissipates more dynamic photons than the Mg atom does, since the Be atom has fewer nucleons than the Mg atom. This implies that nucleons of $Be^{2+}$ are barer than the nucleons of $Mg^{2+}$ in term of the saturation of stationary photons.

In addition, the ionic radii of $Be^{2+}$ and $Mg^{2+}$ are 0.45° A and 0.72° A respectively. $Be^{2+}$ has much shorter ionic radii than $Mg^{2+}$ because the less massive nucleus of $Be^{2+}$ allows its stationary electrons to drift closer to the nucleus.

At a rough estimate, the $Mg^{2+}$ cation is four times larger than the $Be^{2+}$ cation – assuming both of them are spherical in shape, which may not be true. Therefore the Be atom transfers slightly more of its stationary electrons to the oxygen atom than the Mg atom does, since its smaller electron shell can accommodate fewer stationary electrons.

The bonding strength of BeO and MgO are 437 kJ/mol and 363 kJ/mol respectively. BeO has a stronger bonding strength than MgO because the barer nucleons of $Be^{2+}$ with shorter ionic radii exert stronger attraction on the shared electrons within the shared edgorbtoslength with oxygen anion.

In addition, Be and Mg have nine and twenty-three nucleons respectively. More intense exchange of photons among nuclei is needed to break the bond of BeO than the bond of MgO. This may imply a significant role played by nucleons. The presence of more nucleons within the nucleus increase the atom's effectiveness in exchanging photons among nuclei. Therefore, MgO requires a less intense exchange to break its bond because the nucleus of an Mg cation can undergo more effective exchange of photons than a Be cation does.

The edgorbtoslength length of a Be atom becomes shorter after it has reacted with an oxygen atom, but it is still much longer than the edgorbtoslength length of an Mg atom after it has reacted with an oxygen atom. Therefore, the bond between MgO is more brittle than the bond between BeO. The change in edgorbtoslength is due to the modification of its unique nucleus structure as a result of losing stationary photons during the exothermic chemical reaction.

Maybe there is no parallel relationship between the amount of released chemical energy and the chemical bond strength within a chemical compound. Bond strength depends on the uniqueness of the modified unique nucleus structure due to changes in its stockpile of stationary photons. This changes its edgorbtoslengths in term of length, orientation, configuration, and electron shell size. All these changes affect the mutual attraction between the nuclei of a compound. The stationary electrons within their shared edgorbtoslengths in turn determine the actual bonding strength.

Studies of chemistry are not bound by any strict rules, since the experimental outcomes rule over theories. Sometimes we don't know what to expect from a chemical reaction and need to conduct an experiment to ascertain the outcome. We can sometimes predict the outcome, but only for a limited number of elements and compounds. In most cases, we try to reverse-engineer possible chemical reactions based on the actual outcomes of the experiment, in order to figure out what chemical did take place.

There are basically two types of chemical reactions, endothermic and exothermic. Endothermic reactions cause the ambient temperature to drop as a result of heat absorption. Conversely, heat is released to the surroundings during an exothermic reaction.

Since nucleons are much larger than electrons, nucleons are able to stockpile more stationary photons than electrons can during an endothermic reaction. Most dissipating dynamic photons are also derived from the nucleus during an exothermic reaction, because the contribution of photons from stationary electrons must be limited due to their size. Chemical energy is referred to stationary photons that have been stored within nuclei and can be harnessed later, when this stockpile of photons is dissipated from nucleons after a sharing of electrons.

The nucleons of an atom are saturated with two distinct piles of stationary photons. One pile can be replenished after photons have dissipated from the nucleus, whether the dissipation is due to a change in temperature or as a result of a chemical reaction. We have called this group of stationary photons *chemstayphotons*.

Another pile of stationary photons, more abundant than the former, plays an important role in maintaining the stability of the unique nucleus structure. This pile cannot be replenished once its photons are released from the nucleus. We have called this group of stationary photons *nuclstayphotons*. So far, we have only tapped the energy of nuclstayphotons from decaying radioactive substances. Our current technology has not found a feasible way to tap the energy of nuclstayphotons from a stable, non-radioactive substances without having to input a lot of energy.

A new way of sharing electrons takes place after a chemical reaction. A newly formed chemical compound has different chemical and physical characteristics than do the individual reactants before reaction. This clearly attests that the unique nucleus structures of participating reactants before and after the chemical reaction are different from one another. Different stockpiles of stationary photons cause modification of the angular momentum of the nucleons, which leads to different flexing of the unique nucleus structure. This allows modification of their electron shells and edgorbtoslengths, which determine their chemical and physical characteristics. Chemical energy is harnessed from the nucleus rather than from shared electrons.

Naturally an exothermic reaction causes the nucleus structure to lose its stockpile of stationary photons, which normally also causes a reduction of the electron shell's size, especially in metallic atoms. Enlargement of the electron shell occurs in non-metallic atoms even after an exothermic reaction has taken place, as a result of accepting additional stationary electrons contributed from metallic atoms that react with them. Another explanation of the shrinkage in orbital size of metallic atoms and enlargement in orbital size of non-metallic atoms is that the changes could be due to weakening or strengthening of the magnetic energies of the unique nucleus structures after a chemical reaction.

We strongly believe that nucleons of non-metallic atoms transform some of their already weakened angular momentum to boost their magnetic energy. Enlargement, in terms of mosaic of magnetic field strength, of a non-metallic atom's unique nucleus structure helps prop the augmented electron shell. Stationary electrons are also likely to stockpile additional stationary photons from the surroundings to make the enlarged electron shell sustainable as a result of the influence from the augmented mosaic of magnetic field.

The size of an electron shell indirectly determines the nature of the atom: whether it is metallic, metalloid, or insulator. Metallic atoms have very large electron shells. An insulator's electron shell is much smaller. A metalloid's electron shell is slightly larger that of an insulator but smaller than that of a metallic.

There are many different types of metallic atoms. Some metallic materials, like gold, are good electrical conductors but possess low tensile stresses because their electron shells are large. The shells overlap extensively, but their lengths of edgorbtoslengths may be short and don't overlap as extensively. Free electrons flow through such metallic atoms easily. Overlap among shared edgorbtoslength is governed by the nature of the edgorbtoslengths' length and how they overlap, in terms of orientation and configuration.

Iron atoms have rather large electron shells, but their edgorbtoslengths' lengths are much longer than those of gold, and they overlap extensively. An iron atom also has edgorbtoslengths at multiple fronts. Thus, steel, an alloy of iron, is not only a good conductor but also has high tensile strength.

Since most metallic atoms have large nuclei with a large number of nucleons, metallic atoms conduct heat very well. A larger nucleus with more nucleons means more effective and intense photon exchange with neighbouring nuclei.

Ceramics possess a small electron shell as well as very short edgorbtoslengths; therefore, ceramics are not good conductors and are very brittle. Their electron shells are not just small but also do not overlap extensively. Free electrons are forced to unload most of their stationary photons to the surroundings before they emulate the characteristics of the atom's stationary electrons. Therefore, when a current passes through a ceramic, or insulator, it dissipates a tremendous amount of heat. Insulators have high resistance, which hinders the flow of free electrons.

The overlap among a ceramic's short edgorbtoslength is extensive, making the ceramic sturdy. On the other hand, its edgorbtoslengths' lengths are rather short, making it very brittle. In addition, its rigid unique nucleus structure causes it to break easily. Fewer stockpiles of chemstayphotons within its nucleus and a smaller electron shell make it more difficult for its nucleus structure to flex. Thus, ceramic is very brittle.

Exertion of tensile stress on a ceramic acts to pull its atoms apart. This causes distortions in its electron shell and edgorbtoslengths' orientations and configurations. Its unique nucleus structure is not flexible enough to accommodate the changes. Therefore, overstretched ceramic breaks abruptly.

Metalloids fall between metallic and non-metallic elements. Their electron shells are larger than those of non-metallic elements but smaller than those of metallic elements. They are not conductors at room temperature, but they can conduct electricity at higher temperatures. They can also become conductors at higher voltage. High voltage and temperature have a similar effect on them, especially on their stockpiles of stationary photons.

Higher voltage means that there is a source of free electrons that can supply more free electrons per time. If more free electrons flood an array of metalloid atoms, those free electrons that emulate the atom's stationary electrons will dissipate many dynamic photons, which are likely to be absorbed by the nucleons.

Nucleons with larger stockpiles of stationary photons cause enlargement of their electron shell and possibly elongate the length of their edgorbtoslengths too. These changes cause the unique nucleus structure to flex. Higher exchange of photons by the nucleus enables stationary electrons to undergo more intense exchange of photons too. These exchanges lead to further enlargement of the electron shell, and are also likely to further elongate the lengths of their edgorbtoslengths. As a result, electron shells and edgorbtoslengths overlap more extensively than before. These changes allow free electrons to flow more easily.

## Ways to Produce Free Electrons

Solar cells harness sunlight to produce free electrons. Batteries can produce free electrons through chemical processes. A dynamo can transform kinetic energy to electrical energy. Certain materials produce piezoelectricity when they are under compression. We will look at how free electrons are produced by these various methods.

### *Solar Cell*

A solar cell is made from semiconducting materials. Semiconductors streamline the flow direction of freshly produced free electrons.

Certain metallic materials, like selenium, are commonly used in solar cells to help produce free electrons because selenium has photoelectric effects. Selenium strips are placed on the topmost surface of a solar cell to capture dynamic photons from sunlight. Beneath the strips is a semiconductor wafer that helps streamline the flow direction of freshly produced free electrons that do work before being channelled back to the selenium strips.

Selenium atoms have somewhat large electron shells, but both the unique nucleus structure and the stationary electrons have space stockpile even more stationary photons, leading to further enlargement of their already large electron shells upon exposure to sunlight. The overlap among the enlarged electron shells allows freshly transformed free electrons to jump from one electron shell to the next upon exposure to sunlight. Therefore, selenium is said to have photoelectric characteristics.

A nucleus structure that stockpiles a sizeable number of stationary photons allows its stationary electrons to drift farther away. Most likely the specific mosaic of magnetic field of the unique nucleus structure also encourages stationary electrons to stockpile more stationary photons. When the electron shell of a selenium atom undergoes further enlargement, it can transform many of its stationary electrons to free electrons.

We believe the unique nucleus structure of a selenium atom that undergoes a photoelectric effect flexes in a manner that leads to parallel enlargement of its electron shell, followed by shrinkage, followed by enlargement again. This process enables it to catapult freshly produced free electrons away from the atom after they have gained sufficient stationary photons from its surroundings. This is how current is produced by a solar cell.

## Battery

We used to believe that a battery could produce free electrons by transforming chemical energy to electrical energy. We have to agree that through chemical processes on both anode and cathode, freshly produced free electrons flow from anode to cathode. This new understanding debunks the fallacy that a battery transforms chemical energy to electrical energy.

We use a lead-acid battery to illustrate how a battery produces current. A rechargeable battery's anode is spongy Pb, and the cathode is powdered $PbO_2$. The grids are immersed in an electrolyte solution of $\approx 4.5$ m $H_2SO_4$.

Chemical reactions that take place in the anode and cathode are as follows:

Anode (oxidation):
$$Pb(s) + SO_4(aq) \rightarrow PbSO_4(s) + 2e^-$$

Cathode (reduction):
$$PbO_2(s) + 4H^+(aq) + SO_4^{2-}(aq) + 2e^- \rightarrow PbSO_4(s) + 2H_2O(l)$$

Apparently, the anode produces free electrons, which is a push factor. The cathode sucks away any free electrons that have been produced by the anode, which is a pull factor. Push and pull work together to allow freshly produced free electrons to flow from anode to cathode. One can't do without the other in order to streamline the flow of free electrons.

Lead is a soft metal with a large electron shell but its edgorbtoslengths are rather short. Lead atoms slide away from one another because their short edgorbtoslengths do not anchor them well. Its short edgorbtoslengths make lead weak in resisting external tensile forces. Overlap between edgorbtoslengths among lead atoms is not extensive; therefore the atoms have weak bonds, causing them to dislodge from one another easily.

When a lead atom reacts with $SO_4$(aq), it forms $PbSO_4$ and releasing two free electrons in quantum mechanics terms, which is equivalent to twenty free electrons in our unification theory. Lead atom experiences a reduction in electron shell size when it reacts with $SO_4$(aq), after dynamic $SO_4^{2-}$(aq) anion slams against it in an effective collision.

Once the edgorbtoslengths of $SO_4^{2-}$(aq) forge bonds with a lead atom, it releases dynamic photons from its nucleus, then transforms more of its mosaic magnetic field to strengthen the weakened angular momentum of its nucleons. The lead nucleus restructures to a new configuration and orientation. This leads to the reduction of its electron shell size.

Stationary electrons, especially at the edge of the shrinking electron shell of the developing lead cation, easily transform to free electrons by amassing dynamic photons dissipating from the restructuring unique nucleus structure.

A smaller electron shell holds fewer stationary electrons. We are reluctant to say only twenty free electrons are produced for every lead atom that reacts with $SO_4^{2-}$(aq). This is because the actual number of freshly produced free electrons should vary each time a lead atom reacts with $SO_4^{2-}$(aq). The process by which stationary electrons have to amass sufficient stationary photons to transform to free electrons is somewhat random. We don't know exactly how many electrons are freed each time a lead atom reacts with $SO_4^{2-}$(aq). But there is an average number.

While on the cathode, $PbO_2$(s) receives weakened free electrons from the anode. After doing strenuous work, it is likely to be sucked away by the oxygen anions of $PbO_2$(s). We don't know exactly how those weakened free electrons are rationed among the two oxygen atoms. Maybe they share the incoming weakened free electrons equally.

Enlarged oxygen anions are attractive to hydrogen cations that are positively charged. After two hydrogen cations have bonded with the oxygen anions of $PbO_2$(s), somehow the oxygen bond with the Pb is demoted to sub-edgorbtoslength. The weaker bonding with the Pb cation allows newly formed $H_2O$ to break free from the Pb cation.

As soon as the two $H_2O$ molecules detach from the Pb cation, the Pb cation reacts with $SO_4^{2-}$(aq) to form PbSO4(s) through effective collision with dynamic $SO_4^{2-}$.

Recharging a rechargeable battery as soon as possible after discharge prevents damage to both anode and cathode. This is because while the anode and cathode are still hot, PbSO4(s) somehow still makes connection with them due to its enlarged electron shell and hot nuclei, which retain stationary photons and thus retain heat during the discharge process. Once the anode and cathode have cooled down, PbSO4(s) no longer makes a connections with them. This makes recharging them to their status quo condition more difficult. Wear and tear takes place.

## Dynamo

It is widely believed that a dynamo generates electricity by transforming kinetic energy to electrical energy. Can a coil which is an insulator be turned into a permanent magnet or electromagnet that produces electric current? The answer is obviously no. Why must the coil be a conductor like copper?

Copper is a metal with a massive nucleus and a large electron shell; therefore, copper's stationary electrons are at a distance from its nucleus. They can be easily transformed to free electrons after amassing a sufficient number of stationary photons from the surroundings.

The presence of a strong external magnetic field compresses the magnetic field of a stationary electron that is revolving around the copper nucleus. This creates a more compact elastic magnetic field. A more compact elastic magnetic field allows the stationary electron to amass more stationary photons. Therefore, current will be produced as the copper coil cuts the magnetic fluxes.

The number of freshly produced free electrons and how many stockpiles of stationary photons they have depends upon how fast the coil is turning in the magnet and how strong the external magnet is. The faster the coil turns, the more stationary electrons transform to free electrons. A stronger external magnetic field compresses the magnetic field around the stationary electrons harder, which allows more stationary photons from the surroundings to adhere more effectively. A faster-moving coil in a stronger external magnetic field produce an abundance of free electrons embedded with a greater number of stationary photons. These free electrons are more powerful because they have more stationary photons that can be transformed back to dynamic photons to do more work.

A dynamo is one of the most powerful machines for generating ab electrical supply, due to its efficiency in producing powerful free electrons. A dynamo is one of the most efficient machines for transforming mechanical motion into useful output energy: electricity.

Understanding how a dynamo generates electrical current makes it clear that a dynamo does not actually transform kinetic energy to electrical energy. Turning the coil is important in producing electrical current, but the produced current is not transformed kinetic energy. A new understanding of how a strong external magnetic field compresses the magnetic field of metallic stationary electrons before transforming them to free electrons helps debunk the fallacy that kinetic energy can be transformed to electricity.

## Piezoelectric Material

Certain piezoelectric material produce electrical current when it is under compression. The current produced by piezoelectric materials is rather weak because it does not consist of free electrons, but of drifting electrons.

The electron shell of piezoelectric materials is small. When atoms in a piezoelectric material are subjected to a compressive force, their electron shells compress tightly. Stationary electrons are suddenly able to drift from atom to atom. We call these *drifting electrons*, because they are not loaded with stationary photons. Absent a stockpile of stationary photons, the freshly produced current is weak and can hardly do any work.

## Producing Positrons with Electricity-Producing Devices

Solar cells may produce some free positrons in addition to free electrons in their selenium plates. Since the produced current is a direct current (DC), the magnetic field enveloping the outlet wire prohibits any free positrons from flowing in the same direction the free electrons do. Thus, freshly produced free positrons are likely to be trapped in the solar cell.

A rechargeable battery also produces direct current. Freshly produced free positrons are likely to be trapped in the anode.

Both alternating current (AC) and DC dynamos produce sizeable numbers of free positrons, which are always allowed to flow in the opposite direction from free electrons. Since free electrons always outnumber free positrons, it is very difficult to uncover the positrons' existence. Free positrons tend to be saturated with fewer stationary photons than free electrons; this makes them even harder to detect.

Piezoelectric material may produce drifting positrons flowing in the same direction as the drifting electrons, because the produced current is too small and the drifting electrons are not saturated with stationary photons. The flux of drifting electrons per time is not able to produce a significant magnetic field enveloping the circuit they are flowing in. Thus, drifting positrons are not forced to flow in an opposite direction.

Positrons are real, but their existence is hard to detect except through the use of a cloud chamber.

## Cathode Tube

A cathode tube has three distinct parts: a vacuum bulb, a thermionic circuit, and a high-voltage DC supply, as shown in Figure 54.

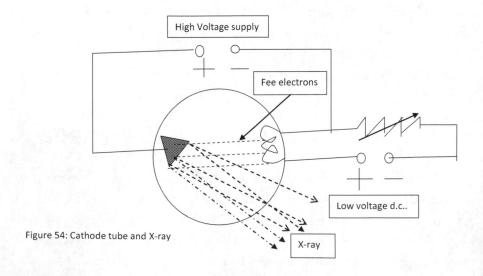

Figure 54: Cathode tube and X-ray

A thermionic circuit consists of a rheostat, a filament, and a low-voltage DC supply. The main purpose of the thermionic circuit is to enable overexcited free electrons of high voltage to jump off a heated filament towards an anode terminal. Tungsten atoms on the heated filament absorb many dynamic photons from the free electrons, which are mainly generated from the low-voltage DC supply. Both the nuclei and the stationary electrons of tungsten atoms stockpile stationary photons from the dissipating heat. This enables the electron shells of the tungsten atoms to enlarge.

The high-voltage DC supply maintains the potential difference between anode and cathode terminals. This allows very compacted free electron with an abundance of stationary photons to be channelled towards the filament in the thermionic circuit. These free electrons jump off from the hot filament towards the anode terminal of the high-voltage supply.

These very compact free electrons from the high-voltage DC supply accelerate towards the anode, due to the high potential voltage difference between cathode and anode. The free electrons are arrested immediately at the anode and suddenly have to emulate the stationary electrons of the tungsten atoms at the anode. This forces the free electrons to dissipate dynamic photons to the surroundings.

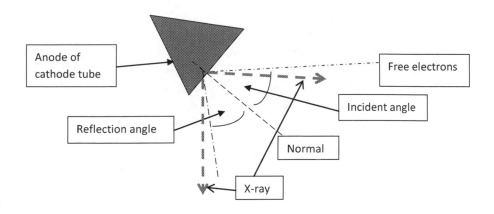

Figure 55: Dispersion of X-ray by the anode of a cathode tube

X-rays scatter widely at a wide angle, as shown by the protruding red arrows. This is because the spurt of free electrons from the thermionic filament is not necessarily travelling in parallel towards the anode. The angle of incidence of all free electrons may not be the same as they hit the anode. The majority have a similar angle of incidence, but don't comply with one of many characteristics of light where the angle of incidence of free electrons is not the same as the scattered angles of the X-ray, as shown in Figure 55. Therefore, the anode's shape is like a wedge at an angle to the direction of most free electrons from the cathode, to facilitate the scattering of X-rays.

This produces high-intensity light with very strong penetrative capability, which we call *X-rays*. X-rays dissipate more dynamic photons per volume per time than ordinary light.

X-rays, with a high intensity of dynamic photons per volume per time, have strong penetrative capability because even after a sizeable number of their dynamic photons have been absorbed, there are still many more dynamic photons that succeed in passing through objects without being captured, especially by nuclei of the object.

## All Electrons Are Identical to One Another

Since all free electrons are enveloped in a strong magnetic field, we expect the free electrons inside a cathode tube to interact strongly with an external magnetic field. The outcome of this interaction obeys Fleming's left-hand rule.

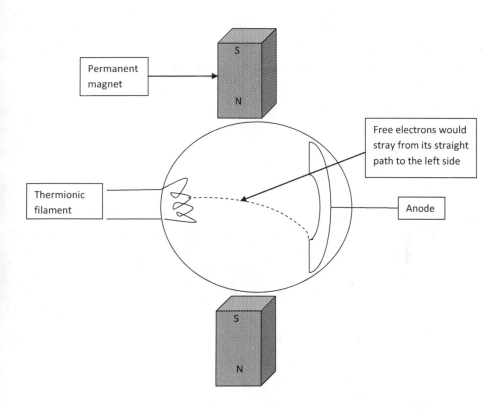

Figure 56: Interaction of free electrons with a fixed external magnetic field

The free electrons that originate from the thermionic filament stray from their usual straight path to the left-hand side, as shown in Figure 56. There is only one spot where all free electrons land on the anode. Therefore we conclude that all free electrons are exactly the same. All of them have the same response after they interact with a fixed external magnetic field.

If there are two types of electrons, then some of those free electrons stray to the left-hand side and others should stray to the right-hand side. Different types of electrons with different intrinsic spins are enveloped by opposite magnetic fields with different orientations and configurations.

Unfortunately, positrons do not end up in a thermionic filament because the thermionic circuit is connected to lower-voltage and higher-voltage DC supplies. All the free electrons that spurt out from the thermionic filament originate from negative polarity. Free electrons from the high-voltage supply are saturated with more stationary photons. Therefore, they are more likely to end up at the outermost edge of the enlarged electron shell, making them more likely to spurt out from the thermionic filament. High-voltage positrons never have a chance to end up in the thermionic filament. Thus, we never sight positrons on the right-hand side.

## Semiconductors

Semiconductors' electron shells are smaller than those of metallic materials, but bigger than those of insulators. The overlap among electron shells is important too in determining whether a material is a good conductor, insulator, or semiconductor. The overlap among the electron shells of semiconductors is not extensive; therefore, semiconductors are not good conductors. But they are better conductors than insulators, whose electron shells barely touch. The electrical conductivity of a semiconductor can be improved by adding traces of impurities, a process called *doping*.

The element used to dope a semiconductor determines its conductivity. Basically, there are two types of extrinsic semiconductors produced by doping: n-type semiconductors and p-type semiconductors. When the element added in doping is "cooked" with the semiconductor, like silicone, its atoms make direct bonds with the semiconductor material.

During heating, nuclei absorb stationary photons. This allow stationary electrons to drift farther away from their nuclei. Stationary electrons that have absorbed many stationary photons can drift farther. Enlarged electron shells overlap extensively. The unique nucleus structure flexes to the new formation. This allow its edgorbtoslengths to re-orient in a new configuration and orientation. Therefore, the bonds among all atoms are different after cooking.

The method of cooling the doped semiconductor determines the overlap among electron shells and the way electron sharing takes place among those atoms. Let's say the cooling-down process is not abrupt. The cooled nucleons eventually lock into a different unique nucleus structure than the one they had before cooking.

Cooling after an intense heating may not enable the unique nucleus structure of some materials to revert back to their status quo orientation and configuration, since their physical characteristics have changed. This clearly shows that the newly formed bonds among edgorbtoslengths shape the unique nucleus structure during the cooling process.

After their electron shell has cooled to ambient temperature, its size shrinks to accommodate the unique nucleus structure. The unique nucleus structure has changed in orientation and configuration. Therefore the edgorbtoslengths orientation and configuration have changed too. Thus, the way atoms share electrons is different as well. They have different physical characteristics before and after the cooking process.

How cooked semiconductors are cooled determines the size of their electron shells, their edgorbtoslengths' orientation and configuration, and the lengths of their edgorbtoslengths' lengths, which in turn determine the extent of overlap of their edgorbtoslengths. All these factors determine the physical and chemical characteristic of the cooked semiconductor.

The way atoms of a mixture of the doped material share electrons determines whether the product is a p-semiconductor or n-semiconductor. Doping silicone with arsenic turns it into an n-semiconductor. Doping silicone with boron makes it into a p-semiconductor. Not only the size of the electron shell is subject to change during doping, but also the extent of overlap between electron shells. This determines the conductivity of a doped material. The level of conductivity determines whether the doped material is a p-semiconductor or n-semiconductor.

Generally, n-semiconductors conduct electricity much better than p-semiconductors. This is because the size of the electron shells in an n-semiconductor's atoms are much larger than those of p-semiconductors. The overlap of electron shells among atoms of an n-semiconductor is also more extensive.

Silicone alone has poor conductivity. Most n-semiconductors and p-semiconductors can conduct much better than silicone.

P-semiconductors have smaller electron shells than n-semiconductors but larger electron shells than silicone. Free electrons that attempt to flow through p-semiconductors must shed their stockpile of stationary photons before they can emulate the stationary electrons of p-semiconductors, which eventually causes free electrons to strand within atoms of p-semiconductors.

Raising the temperature of p-semiconductors enables modification of their conductivity. Dynamic photons flood the stranded free electrons within the p-semiconductor, and the unique nucleus structure flexes after absorbing a substantial number of them. This allows their stationary electrons to drift farther away. The enlargement of the electron shell enhances the p-semiconductor's conductivity as a result of more extensive overlap between the electron shells. Therefore, p-semiconductors' conductivity is heat sensitive, with a rise in temperature enhancing their conductivity.

## Diodes

A p-n junction diode is formed by joining a p-type germanium crystal to an n-type germanium crystal, as shown in Figure 57. The anode is made from the p-type semiconductor, and the cathode is made from the n-type semiconductor. The arrow in the diode represents the direction of current flowing through the diode, as shown in Figure 58.

At low voltage, the diode allows electrons to flow from cathode to anode. N-semiconductors allow free electrons to flood through them, since their electron shells are reasonably large and overlap among them is extensive.

After free electrons have flooded the n-semiconductor extensively, they encounter difficulty in flowing through the p-semiconductor, because the electron shells of the p-semiconductor are smaller and the overlap among its electron shells not as extensive. Many free electrons become stranded in the p-semiconductor. They dissipate a sizeable number of dynamic photons, which in turn heat up the p-semiconductor. The unique nucleus structure of the p-semiconductor flexes after it has absorbed sufficient dynamic photons.

Stationary electrons absorb the dissipated heat too. This leads to enlargement and overlap of the electron shells, which eases the flow of free electrons from cathode to anode.

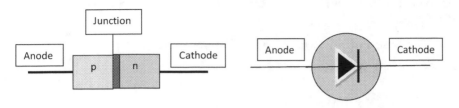

Figure 57: A diode          Figure 58: The schematic diagram of a diode

When free electrons try to flow from anode to cathode in an n-p diode, they are prevented because they dissipate dynamic photons, and the dissipated heat is not well trapped in the diode. Thus it cannot be utilized to heat up the nucleus and stationary electrons of the p-semiconductor. Since the voltage is low, there is not an abundance of free electrons flooding the anode to cause significant heat build-up. Therefore, free electrons are not allowed to flow from anode to cathode of a p-n diode at low voltage.

If the voltage is raised to a certain voltage, however, the free electrons are able to flow. This is because at higher voltage, more free electrons per time flood the p-semiconductor, which leads to significant heat build-up. Heat causes electron shell enlargement and overlap. This allows other free electrons to flow through. Once free electrons have succeeded in flowing through the p-semiconductor, naturally they have no problem flowing through the n-semiconductor.

## Transistors

A transistor is a useful electronic component made from three layers of p-type and n-type semiconductors. A transistor is a three-terminal device consisting of a base (B), collector (C), and emitter (E). There are two types of transistors, n-p-n transistors and p-n-p transistors.

The emitter is much thinner than the collector. In addition, the emitter is heavily doped, whereas the collector is partially doped. The base is a narrow strip that is also partially doped. The reason behind the physical differences among emitter, base, and collector is that the base-emitter interface has a much lower operating voltage than that of the base-collector interface.

Therefore the base-emitter interface is extensively doped to enhance its conductivity.

We will use a p-n-p transistor to explain how free electrons flow through a transistor, as shown in Figure 59.

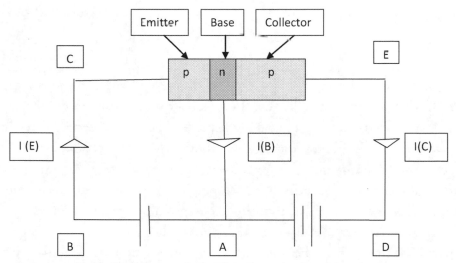

Figure 59: The currents that flow through a transistor

## Base-Emitter Interface

Circuit B-C-Base-A has 1.5 voltage, which is supplied by a 1.5 V battery. The push-pull of this battery ensures free electrons from the negative terminal flow through circuit A-Base-Emitter-C-B before flowing back to its positive terminal. The push factor is supplied by the negative terminal of the 1.5 V battery pumping freshly produced free electrons through circuit A-Base-Emitter-C-B, while the pull factor of the positive terminal of the battery mops up most free electrons after they have done some work.

# Base-Collector Interface

Circuit D-E-Base-A has 3 volts potential difference supplied by two 1.5 V batteries. Freshly produced free electrons flow from the negative terminal through circuit D-E-Collector-Base-A to the positive terminal of the batteries.

Apparently the circuit Base-A has two currents that travel in different directions through it. The free electron flow produced by the negative battery terminals travels through circuit A-Base-C-B. Another, stronger flow of free electrons moves in the opposite direction, travelling through circuit D-E-Base-A to the positive battery terminals. I(B) consists of two currents, namely I(1.5V) and –I(3V). The negative sign on –I(3V) signifies that the current is flowing in the opposite direction.

The segment of circuit of Base-A is like a superhighway that caters for current I(1.5V) and -I(3V). Free electrons from both sources may interact with one another, but fortunately at a minimal rate, because the voltages for both batteries are low.

Since the emitter is short and extensively doped, it readily conducts electricity supplied by the 1.5 V battery through circuit A-Base-Emitter-C-B before returning to the positive terminal of the 1.5 V battery. We expect only a very small number of free electrons supplied by the 1.5 V battery to stray into the collector, because the applied voltage is effective in the circuit of A-Base-C-B. In addition, the collector is likely to be infused with mostly free electrons from the 3 V batteries per time which attempt to flow through the collector. The presence of free electrons in the base, supplied mainly by the 1.5 V battery, helps to heat up the base and the vicinity of the base. This broadens its conductivity boundaries through dissipation of heat from the free electrons flowing from circuit A-Base-C-B back to the positive terminal of the 1.5 V battery. In turn, this has a modifying effect on the conductivity of the collector, which enables a sizeable number of the free electrons that originated from the 3 V batteries to flow through circuit D-E-Base-A to the positive terminals.

If we replace the circuit between the negative terminal of the 1.5 V battery and A with a rheostat to allow changes to the intensity of the current originating from the negative terminal of the battery, we witness parallel changes in the current that eventually flows through circuit D-E-Base-A. The different intensity of current flows from the 1.5 V battery have different heating intensity that affects the base of the transistor and its vicinity (especially the collector). This modifies the conductivity of the collector, which in turn allows different intensity of current to flow through circuit D-E-Base-A. If a transistor is properly based, the collector current or output current is directly proportional to the base or input current, and the transistor acts as a current amplifier.

## Radio

A radio can tune in to a particular channel or broadcast with the aid of an LC circuit. Basically, an LC circuit consists of a solenoid with specific turns and a variable capacitor, as shown in Figure 60.

A long wire antenna and solenoid are the instruments necessary to produce sufficient free electrons as a result of interaction with a specific radio wave broadcast by a specific radio station. A radio wave dissipates dynamic photons with a specific configuration and saturation per time. The dynamic photons are intercepted by stationary electrons in the metallic atoms of the antenna and solenoid. The stationary electrons transform to free electrons after they have amassed sufficient stationary photons per time.

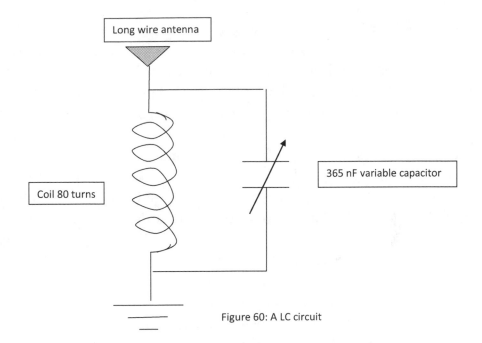

Figure 60: A LC circuit

The capacitor acts like a bucket that fills up with freshly produced free electrons from the antenna and solenoid. Its "brim" is on one side of its plate. When full, it attempts to empty itself before allowing the other plate of the capacitor to fill with free electrons to its brim again. This enables the capacitor to produce an oscillating alternating current.

A variable capacitor is used to fine tune a radio to a particular channel. If the radio is moved to a new location, then the variable capacitor must be adjusted in parallel to the intensity of the radio wave in the new location. At a greater distance from the radio station, the intensity of the radio wave is reduced. Therefore the radio wave has a lower number of dynamic photons per volume per time. Thus, the antenna and solenoid produce fewer free electrons, and a smaller bucket is required to attain resonance in the LC circuit. A smaller bucket requires less time to fill; therefore, a smaller bucket has a higher frequency of topping and emptying. At a greater distance from a radio station, we normally tune the radio to a higher frequency. But this is a by-product of how the capacitor functions. It doesn't imply that the radio waves emitted from the station have changed frequency.

It is a misconception that a more powerful electromagnetic wave has a higher frequency. That misconception comes from the belief that EMWs are waves, which as we have shown is wrong. EMWs are dissipating dynamic photons per volume per time. Even if the intensity of dynamic photons per volume per time can be the same, different distribution causes variations in terms of freshly produced free electrons from a specific antenna and its solenoid after interaction of the radio wave with the LC circuit. Therefore, the operating bandwidth of a specific telecommunications company can cater to many clients with different telephone numbers, not just because of the use of different frequencies in the bandwidth but also because the operating EMWs have distinctive distributions of dynamic photons per volume per time.

A cell phone is different from a radio because a cell phone always keeps its identification number, which is its phone number. Each cell phone in a single cell is identified according to its phone number. Therefore, each cell must have an individual cell phone transmission station. If a cell phone operates like a radio, the phone is required to adjust its LC circuit to keep connected. Then how can the telecommunications company keep tabs on so many cell phone users at any one time in different locations? Surely there will be confusion about how all those cell phones should be connected to the transmission station.

A cell phone uses two different frequencies for receiving and transmitting. If it used a single frequency for both transmission and reception, it would operate like a walkie-talkie, which can only transmit or only receive at any given time. One cannot talk and listen at the same time.

Radio waves are not mechanical waves, but a specific range of dynamic photons per volume per time. More intense radio waves have a higher number of dynamic photons per volume per time than weaker radio waves. If we understand how a radio tunes into a particular broadcast, then we understand how a cell phone operates.

Let's say the antenna and solenoid of a radio and its tuned variable capacitor "listen" for a particular channel and are located at a specific distance from the radio station. The antenna and solenoid tap dissipating dynamic photons emitted from the radio station and produce a range of specific free electrons.

Radio signals with different variations in term of saturation of dynamic photons per volume per time, which have been symbolized by the different thickness of each strip with each of them has been distanced apart by a duration of 1/(2T)s. Every time the radio signal sweeps through the LC circuit, subsequently a specific range of free electrons would be produced in parallel to the saturation of the radio signals themselves. The arrow shows the dissipating direction of the radio signals away from a particular radio station towards a radio.

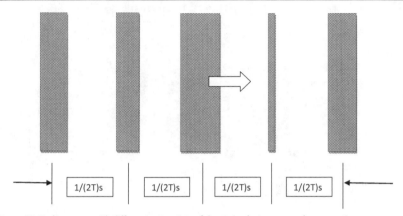

Figure 61: Radio waves with different saturation of dynamic photons per volume per time

Initially, the oscillation of freshly produced current in the tuned LC circuit of a radio is exactly like the operating frequency of the particular radio station. Radio signals are clusters of dynamic photons per volume per time with different intensity that sweep through a region. Each individual cluster of dynamic photons is separated by a time duration of 1/(2T)s. The cluster sweeps through the LC circuit of a radio, as shown in Figure 61. Its frequency is exactly the same as the oscillating frequency of the LC circuit of the radio.

Each time a cluster of dynamic photons sweeps through the LC circuit, a specific range of free electrons is produced. Free electrons result from the interaction of the radio signal with the antenna and the solenoid of the LC circuit, in parallel with the intensity of the radio waves. The dynamic photons of the radio waves tend to adhere to the stationary electrons in the radio. When sufficient stationary photons have been amassed, the stationary electrons transform to free electrons. The more intense the radio wave, the more free electrons are produced. Of course, some dynamic photons from the radio waves are absorbed by nuclei in the radio.

Very soon the oscillating flow of freshly produced free electrons in the LC circuit becomes coherent and orderly. One plate of the capacitor is flooded with freshly produced free electrons at a time. It attains full resonance when the recharging and discharging frequencies of the capacitor are in sync with the operating frequency of a radio station. Then the radio is said to be "tuned in" to that particular channel.

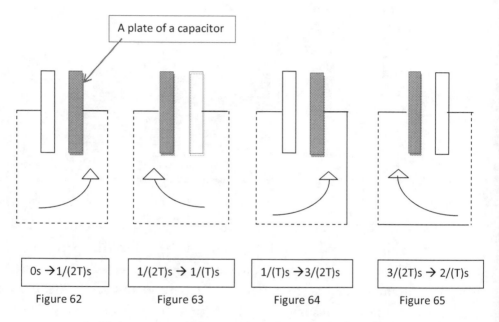

<table>
<tr><td>0s →1/(2T)s</td><td>1/(2T)s → 1/(T)s</td><td>1/(T)s →3/(2T)s</td><td>3/(2T)s → 2/(T)s</td></tr>
<tr><td>Figure 62</td><td>Figure 63</td><td>Figure 64</td><td>Figure 65</td></tr>
</table>

Oscillating current within a capacitor of a LC circuit of a radio

The freshly produced free electrons start to saturate either one of the plates of the capacitor. As more free electrons cluster on the plate, the tension among them increases. Similarly, the dielectric material sandwiched between the plates flexes its nucleus structure according to the saturation of free electrons in the plate. The dielectric materials temporarily holds up the free electrons before "transforming" them back to free electrons after their nucleus structure flexes back to status quo orientation and configuration. This causes free electrons to flow from one plate to the other plate. One plate is saturated first, then diverts the free electrons to the other plate.

The flow direction of free electrons is indicated by the arrows shown in a sequence from Figure 62 through Figure 65.

This process goes on as long as the radio is operating. Some free electrons cannot be recycled because they convert to stationary electrons by dissipating most of their stationary photons to the surroundings. But most freshly produced free electrons keep oscillating. The duration of saturation of the capacitor plates is illustrated in the sequence of Figure 62 through Figure 65. The frequency of the tuned radio signal is said to be 1/Ts.

The tuned capacitor times the duration of the oscillating current, especially when free electrons pass through the solenoids while a radio wave is passing through them. When the atoms of the solenoids are saturated with a sizeable number of free electrons diverted from the capacitor, their electron shells are augmented. When a radio wave passes through the solenoids at this instant, it is more effective in spawning additional free electrons.

Similarly, some stationary electrons within the enlarging electron shell can transform to free electrons easily by absorbing dynamic photons from the passing radio wave. The condition of sync timing occurs when remnant free electrons pass through the solenoids at the moment when a radio wave is sweeping through the solenoids. This enables production of a significant number of free electrons, and is called *resonance*.

When free electrons diverted from a capacitor plate pass through the solenoids of an LC circuit while a radio wave is sweeping through, free electrons envelop the solenoids with their superposed magnetic fields. Freshly produced free electrons are guided to flow in the same direction of existing free electrons. The streamlined flow is directed such that only one of the plates of a capacitor is filled up at any one time. Thus, oscillating current flows to and fro within a capacitor.

Let's say there are two competing channels, one with a higher operating frequency and the other with a lower operating frequency. What happens to the flow of freshly produced current in the LC circuit that is already set to a particular channel when it reacts with these two competing radio signals?

Higher-frequency EMWs cut through the antenna and solenoid more often. Before the remnant free electrons can be diverted through the solenoids from one plate of the capacitor, the radio waves have already swept through the solenoids, producing a minuscule number of free electrons. No resonance takes place. Absent a streamlined flow, these few free electrons scatter haphazardly. Therefore, they do not help to shape the resonant current in a capacitor.

A tuned capacitor that attempts to attain new resonance with the higher-frequency radio signals will fail to do so. As soon as one side of the plate of a tuned capacitor discharges its stored free electrons, a small amount of freshly produced haphazard current from the higher-frequency radio wave makes its way towards the plate. This is unlikely to dampen the smooth flow of current in the tuned LC circuit.

Radio waves of lower frequency take longer between sweeps through the LC circuit. Again, they are out of sync and produce few free electrons. Thus, a lower-frequency radio waves cannot establish resonance with an LC circuit tuned to a higher-frequency radio wave.

Freshly produced free electrons, whether from tuned-in radio waves or other radio waves, interact among themselves to ensure that the selected radio frequency attains resonance. Every time the tuned radio waves cut through the LC circuit, the maximum number of free electrons is produced in parallel to the wave's intensity in terms of saturation of dynamic photons per volume per time. When other radio signals cut through the LC circuit, many fewer free electrons are produced. They are not effective in producing a streamlined flow.

Maybe the resonant production of free electrons by the LC circuit affects the absorption of heat in the LC circuit and enables maximum production of free electrons when the selected radio waves are sweeping through the LC circuit.

The LC circuit is tuned in to a specific frequency of radio wave so that the LC circuit attains resonance, producing the maximum possible number of free electrons each time a particular radio wave sweeps through the solenoid. The capability of an LC circuit to tune in to a particular channel enables our various gadgets – radio, TV, cell phone, Internet modem, remote control equipment, satellite communication, and so forth – to operating meaningfully.

Resonant current produced by the LC circuit passes through a transistor, which functions as an amplifier. It converts the radio's output to a meaningful audio message via a loudspeaker. This is how radio works.

# Magnetron and Microwave Oven

Microwave ovens operate at a very high voltage, up to 2,000 volts. The magnetron is the heart of a microwave oven. The emitter is where a sizeable number of dynamic photons spurt out, derived from the free electrons emitted from the magnetron.

The setting of a magnetron is like that of a cathode tube. There is a heating cathode filament that is connected to the high-voltage DC supply. In addition, there is an anode block with several resonant cavities in it.

The distinctive different between a cathode tube and the magnetron is that there is a strong magnet at the base of the magnetron. The main purpose of the magnet is to prevent free electrons emitted from the cathode filament from dissipating their stockpile of stationary photons on their way to interacting with a strong external magnetic field. The free electrons' magnetic fields are suppressed, creating a more compact elastic magnetic field that allows the free electrons to retain as many stationary photons as possible.

The presence of a strong external magnetic field at the base of the magnetron also causes the free electrons to spiral out towards the anode block. They are likely to retain most of their stockpile of stationary photons once they embed in the anode block.

If not for the presence of a strong external magnetic field, the free electrons emitted from the cathode filament would be likely to shed their stationary photons and form X-ray radiation once they hit the anode block. The magnetron is designed to flood the anode block with high-tension free electrons. Free electrons produced at a much higher voltage can be saturated with more stationary photons before being emitted from the cathode filament. A strong external magnetic field interacts with those spiralling, high-tension free electrons to ensure they retain as many stationary photons as possible before they arrive at the anode block. In addition, the spiralling motion of the high-tension free electrons enables them to be absorbed into the anode block, ensuring that they remain very compact and with an abundance of stationary photons.

High-tension free electrons are tapped from the anode block by an output coupling loop. A copper rod connects the anode block to the emitter.

The electricity supplied to the magnetron is high-tension direct current generated from alternating current. Therefore, the supply of high-tension free electrons from the cathode filament is emitted in pulses.

The high-tension free electrons that have been tapped from the anode block rush along the copper rod to the emitter. Once they reach the dead-end emitter, they quickly unload stationary photons, transforming them to dynamic photons. The free electrons then emulate the stationary electrons in the emitter. The build-up of stationary electrons at the emitter is mopped away by the anode.

A microwave emitted by a magnetron has a high intensity of dynamic photons per volume per time. Such intense heat can be used to cook foods.

Beside the microwaves, magnetrons are also used to power radar. The pulses of high-tension free electrons produced by a magnetron are channelled towards a radar-emitting antenna, to be scattered at a high intensity of dynamic photons per volume per time. As with a microwave, the pulses of high-tension free electrons rush to the dead end of the radar-emitting antenna. There they unload most of their stationary photons and transform to stationary electrons.

Some dissipating dynamic photons are absorbed by, say, an aircraft and are deflected back again. If significant deflected dynamic photons are intercepted by the radar receiver, it converts stationary electrons to free electrons. The freshly produced pulse of free electrons is recognized as the presence of an aircraft in the airspace the radar receiver is pointing to.

The predicted distance between an aircraft and the radar is calculated based on the lapse duration for the dynamic photons to dissipate from the radar emitter to the aircraft before being deflected from the aircraft back to the radar receiver.

Understanding how a microwave oven cooks foods or a radar detects aircraft reinforces our belief that all electromagnetic waves are actually dissipating dynamic photons per volume per time. EMWs are not waves.

# Lightning

Some clouds don't collect any charge at all. Some clouds are positively charged and others negatively charged. Lightning takes place when a cloud attempts to unload its excess charge.

Water molecules have dipole characteristics. An oxygen anion is negatively charged, while hydrogen cations are positively charged. Hydrogen atoms lose some of their stationary electrons to the oxygen atom of a water molecule, so an oxygen anion is negatively charged because it has additional stationary electrons. A hydrogen cations is positively charged by losing some stationary electrons to the oxygen anion. This is how water molecules acquire their dipole characteristics.

At different altitudes, the dipole characteristics of water molecules change in parallel to localized gravitational potential energy, ambient pressure, and temperature.

At different altitudes, the stockpiles of stationary photons on both oxygen anions and hydrogen cations are subjected to changes in parallel to their magnitude in terms of localized gravitational potential energy. The nuclei of the oxygen anions flex in a certain manner due to changes to the strength of their nucleons' angular momentum, which in turn make changes to the way it shares stationary electrons with hydrogen cations. Different ambient pressure and temperature at a particular altitude also determine the stockpiles of stationary photons on their stationary electrons, which in turn determine the size of their electron shells and the length of their edgorbtoslengths.

Differences in the stockpiles of stationary photons in a unique nucleus structure determine the flexing of the unique nucleus structure, which in turn causes modifications to the magnitude of the dipole of the water molecule at different altitudes. The size of their electron shells and the length of their edgorbtoslengths determine how electron sharing takes place between air molecules and water molecules in clouds at a particular altitude.

Air molecules at different altitudes have different stockpiles of stationary photons in their nuclei due to the effects of localized gravitational potential energy. This also modifies the flexure of their unique nucleus structure, which in turn modifies their edgorbtoslengths in terms of orientation, configuration, and length. Sometimes, localized gravitational potential energy, ambient pressure and temperature also determine the stockpiles of stationary photons within the stationary electrons.

Denser stationary electrons that are saturated with stationary photons can drift farther away from the nucleus. Lighter stationary electrons with fewer stationary photons drift closer to the nucleus. Localized gravitational potential energy determines the ways ionization occurs in air molecules attached to water molecules in the crystalline structures of a rain cloud.

Hydrogen cations of a water molecule have a limited number of stationary electrons. The oxygen anion of a water molecule is likely to retain its excess stationary electrons. Research on rainwater shows that when a lightning process has taken place, it does not in any way lead to changes to the chemical characteristics of rainwater. In other words, water molecules are rather inert. Therefore, the excess free electrons accumulated by a rain cloud are not derived from the water molecules in the rain cloud. We suspect the build-up of charges on a rain cloud must derive from air molecules brushing against the air molecules that adhere to the crystalline structure of water molecules in the rain cloud.

The earth's atmosphere is 70 per cent nitrogen and 20 per cent oxygen. Nitrogen gas can have several oxidation numbers, namely +5, +4, +3, +2, +1, -1, -2, and -3. Oxygen can only have one oxidation number, -2. Therefore, oxygen molecules that adhere to the crystalline structure of a rain cloud can only retain excess free electrons after other air molecules brush against it.

Nitrogen molecules that adhere to a crystalline structure can accumulate or lose excess stationary electrons depending on the altitude. Localized gravitational energy, ambient pressure, and temperature are different at different altitudes, which affects the way nitrogen molecules share electrons with water molecules in the crystalline structure of the rain cloud.

Charges build up in a rain cloud when weather conditions are windy. Gusting air rubs against the trapped air molecules that adhere to the crystalline structure of the rain clouds. The friction allows them to either accumulate or lose stationary electrons depending on the altitude. The crystalline structure of the cloud acts like the dielectric material of a capacitor, temporarily holding up any excess stationary electrons in the air molecules that adhere to the cloud.

When raindrops precipitate from the cloud, lightning ensues. The air molecules can no longer hold their excess stationary electrons, which transform to free electrons. The freshly produced free electrons dissipate a sizeable number of dynamic photons to their surroundings. This heats the cloud, melting some of its icy crystalline structure. Melting triggers more precipitation, which leads to even more free electrons being produced. A tipping point is reached, and in a split second, lightning occurs.

A rain cloud that is saturated with stationary electrons transforms them to free electrons when raindrops start to precipitate. Free electrons from the precipitating rain cloud flow to something positively charged, like another rain cloud. Lightning then jumps from the negatively charged rain cloud to the positively charged rain cloud.

More often than not, lightning jumps from negatively charged rain clouds to the earth. The earth is constantly hungry for more stationary electrons, as a result of the expansion of the universe. This link may not be obvious at first. Since the solar system is drifting in parallel with the direction of expansion of the universe, its atoms have an uncanny capacity to stockpile stationary photons. Complete darkness envelops a deserted area on a moonless night because atoms are so hungry for stationary photons that they absorb more dynamic photons from the surroundings than they emit.

A unique nucleus structure that increases its stockpile of stationary photons increases its mass. This enables its electron shell to hold additional stationary electrons, as it has increased its gravitational clout. Thus, the earth constantly craves more stationary electrons. The earth is always positively charged. As a result, lightning discharges from negatively charged rain clouds down to the earth.

Lightning is intensely bright, which implies that stationary electrons at higher altitudes can amass a lot of stationary photons due to high localized gravitational potential energy at higher altitude. Therefore, air molecules at higher altitudes can be ionized easily.

# GLOSSARY

**Accretion disk:** A spheroid planar that consists of celestial bodies revolving around a black hole or neutron star at very high speeds under the influence of the spinning-gravitational effects of its superstructure.

**Angular momentum:** An acting force on a particle. It is obvious that electrons possess angular momentum, as they always revolve around the nucleus. Protons and neutrons also possess angular momentum. Since nucleons arrange themselves in a unique nucleus structure in which their angular momentums balance out, their possession of angular momentum is not that obvious. Photons, electrons, protons, and neutrons derive their angular momentum from kinetic energy released during the Big Bang explosion. Increasing the stockpile of stationary photons on a nucleon or electron strengthens its angular momentum. As the universe continues to expand, more angular momentum is transformed to universal gravitational potential energy. Thus, formerly stable elements eventually become radioactive substances as the angular momentum of their nucleons weakens.

**Black hole:** A collapsed, very massive star. When a very massive star is running out of nuclear fuel, it collapses under its own weight and undergoes a supernova explosion, during which its atoms are reduced to preons. Preons are the finest form of subparticles. Positive preons are surrounded with negative preons and vice versa, forming a very dense, sturdy structure called a superstructure. The superstructure spins at a very fast speed and is enveloped be a very strong magnetic field. Upon completion of this process, a black hole is said to have born.

**Bondbital:**   Another name for edgorbtoslength. The point where an electron-sharing process is likely to take place between atoms. The orientations and configurations of bondbitals are moulded by an atom's unique nucleus structure.

**Centre *or* core:**   The zone where the Big Bang explosion takes place. All celestial bodies are heading out from the centre of the universe as a result of the Big Bang. All atoms in the core are saturated with stationary photons. They have very high kinetic energy and are surrounded by very strong magnetic energy. All celestial bodies in the core are extremely hot during the Big Bang and shortly thereafter. Therefore, no life forms exist in the core.

**Chemical energy:**   The chemical characteristics of an atom are different before and after a chemical reaction. There are two distinct types of chemical reactions, endothermic and exothermic. Since nucleons are much larger than stationary electrons, this makes us believe that nucleons play an important role in contributing dynamic photons during an exothermic reaction. Large nucleons are likewise able to mop up dynamic photons during an endothermic reaction. Changes in the stockpiles of stationary photons on the nucleons either weaken or strengthen their angular momentum, which leads to modification of their unique nuclear structure and their edgorbtoslengths. The unique nucleus structure also flexes differently. Therefore the manner in which electrons are shared between atoms is different before and after a chemical reaction. Its physical and chemical characteristics change. A big portion of chemical energy is harnessed from the nucleus during a chemical reaction, rather than from electrons.

**Chemstayphotons:**   This neologism refers to a stockpile of stationary photons that adhere to the nucleons of an atom and can be replenished after they have been released due to changes in temperature or as a result of a chemical reaction. Chemical energy is the energy that is tapped from chemstayphotons during a chemical reaction.

**Dense radioactive substances:** A range of radioisotopes from actinium to elements with atomic number 129, likely to undergo proton or neutron emissions during decay. The presence of hydrogen gas in the atmosphere of a celestial body indirectly signals that its core used to be composed of dense radioactive substances. They dissipate intense heat during their decay, but considerably less intense than that of very dense radioactive substances. They transform to light radioactive substances, very light radioactive substances, or both upon decay.

**Drifting electrons:** Formerly stationary electrons that suddenly drift from atom to atom due to the presence of compressive forces against a piezoelectric material. Electron shells among such atoms overlap each other extensively, allowing stationary electrons to drift from shell to shell. A drifting electron has a stockpile of stationary photons similar to that of a stationary electron. Therefore, they cannot do strenuous work.

**Dynamic photons:** Photons that move freely on their own. Photons that adhere to electrons and nucleons are called *stationary photons*.

**Electron shell:** The region around an atomic nucleus in which electrons have their orbits. Every atom has one electron shell. The geometrical shape and size of an electron shell is moulded by the atom's unique nucleus structure. The thickness of the electron shell varies from location to location in parallel with the uniqueness of the unique nucleus structure. The electron shell is supported by the mosaic of magnetic field of the nucleus.

**Edgorbtoslength:** An acronym of the phrase *edge* of *orb*ital *to* electron *s*hell *length*. Another name for edgorbtoslength is *bondbital*. It is the location where an electron-sharing process is likely to take place between atoms. The unique nucleus structure moulds the edgorbtoslength. Stationary electrons frequent the space above the portions of the unique nucleus structure where protons are more saturated. This is where an electron-sharing process is likely to occur. The bond between atoms is stronger if the edgorbtoslength is longer. Shared stationary electrons can sweep past wider, longer spaces without crashing against each other frequently.

Bonds between atoms are stronger if there are more stationary electrons within the shared edgorbtoslengths.

**Free electrons:**   Electrons that have gained sufficient stationary photons to enable them to drift from atom to atom, due to their greater mass and stronger angular momentum.

**Ghost photons:**   Imploding photons from the edge of the universe after they have converted all their kinetic energy to universal gravitational potential energy. The existence of ghost photons gives rise to cosmic background radiation.

**Interlocking attraction:**   A neutron consists of electrons, a proton and some quarks. A proton consists of a strong, positively charged quark and a weak, negatively charged quark. Electrons of a neutron are attracted to the positive quark of a proton. Likewise, the proton of a neutron is attracted to the negative quark of a proton. Therefore, neutrons within a nucleus provide interlocking attraction to its protons in order to maintain the cohesiveness of the unique nucleus structure.

**Hypergiant stars:**   Stars that are more than a million times brighter than the sun. They are likely to turn into black holes, regardless of whether they are embedded with dense radioactive substances, very dense radioactive substances, or both.

**Intrinsic spin:**   All charged particles, namely photons, electrons, protons, and neutrons, possess intrinsic spin. The spinning of charged particles gives rise to the magnetic fields that envelop them. When dynamic photons transform to stationary photons, they hasten their intrinsic spin, tapped from kinetic energy (angular momentum). This produces a stronger magnetic field to envelop them. Therefore, free electrons saturated with stationary photons are enveloped in a stronger magnetic field.

**LC circuit:**   A parallel circuit that consists of a variable capacitor and a solenoid, enabling a radio to tune in to a particular channel.

**Light radioactive substances:**   Radioisotopes ranging from niobium to radium. They are likely to transform to heavy metallic elements and dissipate a moderate amount of heat.

**Localized gravitational potential energy:**   Localized gravitational potential energy has an effect on atoms and molecules at a given altitude. It results from the gravitational effect of the celestial body on the unique nucleus structure of the atom, causing it to flex in a certain orientation and configuration. Flexure can cause the atom to stockpile stationary photons more effectively from the surroundings, and can also cause it to intensify its exchange of photons with its surroundings. An atom's stationary electrons may become easier to ionize at a certain altitude. The earth's localized gravitational potential energy gives rise to lightning, Van Allen belts, and auroras.

**Main sequence:**   Most stars in the universe are in the main sequence. Such stars do not end their lives in a violent explosion, or supernova, because they are not massive enough or embedded with mostly dense radioactive substances. They turn into red dwarfs as their nuclear fuel depletes, then cease to emit light and heat.

**Middle rim:**   The zone in which all celestial bodies have converted about half of their kinetic energy to universal gravitational potential energy. The expansion of the universe is much slower now. Change in physical and chemical characteristics of chemical compounds is also slow, as well as more sustainable and predictable. This gives rise to the existence of life forms. All habitable planets in the middle rim are teeming with life forms.

**Monomegastar:**   A humongous star that is usually found at the heart of a massive galaxy. The entire galaxy gyrates in sync with the movement of the monomegastar. When the monomegastar ages, it turns into a black hole because it collapses under its tremendous weight.

**Mosaic of magnetic field:**   A mosaic of magnetic field consists of a mixture of north and south magnetic polarities in a random arrangement in parallel to the uniqueness of the unique nucleus structure. It supports stationary electrons to enable them to surf on top of it. The presence of mosaic of magnetic field of the nucleus prevents stationary electrons from caving in towards the nucleus when atoms rest on top of one another.

**Neutron star:**   A black hole lookalike whose superstructure consists of quarks rather than preons. Its lesser mass does not compress as hard as a black hole's does when it is collapsing under its own weight.

**Nuclstayphotons:**   A neologism referring to a stockpile of stationary photons that adheres to the nucleon of an atom and plays an important role in maintaining the stability of its unique nucleus structure. Nuclstayphotons cannot be replenished after they have been released from the nucleus. So far, we have only managed to tap energy from nuclstayphotons as part of the decay of radioactive substances. We call this *nuclear energy*. Our current technology has not yet found a feasible way to tap nuclear energy from a stable, non-radioactive substances without having to input a lot of energy.

**Outer rim:**   A zone in which celestial bodies have converted almost all their kinetic energy to universal gravitational potential energy, before the universe has expanded to its maximum size. Most outer-rim stars have entered their aging state. Monomegastars in the outer rim have likely turned into black holes. Life forms cease to exist in the outer rim because of low kinetic energy conditions.

**Relativistic Electron-Proton Telescope (REPT):**   An instrument that measures the emitted radiation of electrons, hydrogen atoms, and other gaseous atoms and molecules at very high altitudes. The REPT that measures protons is shielded with a thin lead plate, whereas the REPT that measures electrons is shielded with a glass plate.

**Spinning-gravitational effect:**   Any celestial body that revolves on its own rotational axis has a spinning-gravitational effect on objects that are attracted to it. Due to the spinning-gravitational effect, such objects won't nosedive towards the spinning celestial body. Instead, they revolve around it. Therefore, we cannot treat any spinning celestial body as a point mass when we examine its relationship with objects that are attracted to it. The existence of the spinning-gravitational effect assures us that the existence of exoplanets is rather common in the universe.

**Stationary electrons:**   Electrons that are confined within an atom. They have fewer stockpiles of stationary photons as compared to free electrons.

**Stationary photons:**   Photons that adhere to electrons, protons, and neutrons. Dynamic photons don't lose their energy after they transform to stationary photons. Part of their kinetic energy (angular momentum) is transformed to magnetic energy, hastening their intrinsic spin.

**Sunspots:**   Sunspots are born near the magnetic poles of the sun. They are patches of ferromagnetic material that turn into a stronger magnet under the presence of a strong external magnetic field. Stronger ferromagnetic material has a bigger appetite for stationary photons. As it amasses more stationary photons, it turns into a much stronger magnet which in turn can amass even more stationary photons. As sunspots drift closer to the equator of the sun, the weaker external magnetic field cannot help sunspots hold their stockpiles of stationary photons. The stationary photons are released as dynamic photons. Subsequently, sunspots become weaker magnets. This triggers a further release of dynamic photons, and so on. When a large release of dynamic photons takes place, a solar flare occurs. Part of the sunspot's ferromagnetic material is flung into the air before falling back onto the sun again. The released dynamic photons assault the space facing the solar flare.

**Supergiant stars:**   Stars that are more than ten thousand times the luminosity of the sun. Their cores may be filled with dense radioactive substances, very dense radioactive substances, or both. When they collapse under their own weight as they are running out of nuclear fuel, they turn into either a black hole or a neutron star, depending on their massiveness.

**Super monomegastar:**   A black hole continues to gobble up other celestial bodies to fuel the construction of its superstructure. During implosion, preons transform their universal gravitational potential energy to kinetic energy (angular momentum), and heighten their intrinsic spin, which strengthens their magnetic energy. The continuous build-up of magnetic energy among quarks frees them. They then transform to electrons, neutrons, and protons, eventually forming atoms. The mass of such a body turns out to be much greater than a monomegastar. Therefore, we call it a super monomegastar.

**Superstructure:** When a monomegastar collapses under its own weight, atoms crunch into quarks or preons before fuelling the formation of a superstructure. The superstructure is a tremendously dense lump of matter made out of quarks or preons. Positive quarks or preons are surrounded by negative quarks or preons and vice versa. A superstructure is usually found in the heart of a black hole. It is surrounded with a very strong magnetic field and spins at a very high speed because it has conserved the angular momentum of its predecessor atoms. A neutron star has a less pure superstructure consisting of quarks.

**Type A stars:** Stars with surface temperatures between 8,000 and 10,000 kelvins.

**Type B stars:** Stars with surface temperatures between 12,000 and 25,000 kelvins.

**Type F stars:** Stars with surface temperatures between 6,000 and 7,500 kelvins.

**Type G stars:** Stars with surface temperatures between 5,000 to 6,000 kelvins. The sun is categorized as a type G star.

**Type K stars:** Stars with surface temperatures between 3,000 and 4,000 kelvins.

**Type M stars:** Stars with a surface temperature of approximately 3,000 kelvins.

**Type N stars:** The very first form of stars, embedded with non-radioactive substances. Their nuclei are enriched with an abundance of stationary photons.

**Type O stars:**   Stars with surface temperatures between 30,000 and 40,000 kelvins.

**Unique nucleus structure:**   A unique arrangement of nucleons within the nucleus of an atom. Protons will not sit close to other protons, so neutrons diffuse the tension among protons within the nucleus. The nucleons of an element arrange themselves in a structure unique to that element. A stable unique nucleus structure must be harmonized in terms of gravitational, electrostatic, and magnetic energies among all its nucleons. In addition, all its nucleons' angular momentum must be strong enough to oppose the repulsive forces among its protons. Changes in the stockpiles of stationary photons within the nucleus have a modifying effect on the strength of the angular momentum of the nucleons, which leads to a different flexure of the unique nucleus structure. The unique nucleus structure also determines the configuration and orientation of its edgorbtoslengths, which in turn determine the physical and chemical characteristics of its chemical compounds.

**Universal gravitational potential energy:**   All other energies, namely kinetic, intrinsic spin, magnetic, and angular momentum, are derived from universal gravitational potential energy. As the universe continues to expand, more of its kinetic, intrinsic spin, magnetic, and angular momentum energy is gradually transformed to universal gravitational potential energy. The reverse process takes place during the implosion of the universe after it has reached to its maximum size.

**Van Allen belts:**   A region in near space where gaseous atoms and molecules can effectively stockpile stationary photons from their surroundings due to their localized gravitational potential energy. Those gaseous atoms and molecules are loaded with stationary photons, and they unload them onto any objects that collide against them. Therefore, Van Allen belts pose a threat to astronauts, satellites, and spacecraft.

**Very dense radioactive substances:**   A range of radioisotopes with atomic number 130 and above. They are likely to undergo alpha emission during radioactive decay. Detection of helium gas in the atmosphere of a celestial body is a signal that its core used to be embedded with very dense radioactive substances. When very dense radioactive substances decay, they dissipate many more dynamic photons to the surroundings than dense radioactive substances. Very dense radioactive substances are more unstable than dense radioactive substances; therefore their half-lives are much shorter. They are likely to transform to very light, light, or dense radioactive substances during their decay.

**Very light radioactive substances:**   Radioisotopes ranging from lithium to zirconium. They are likely to transform to non-metallic or light metallic elements, and dissipate a very low intensity of heat.

**White dwarf:**   A bright star that is not very massive. It is likely to undergo a type Ia supernova explosion when it is collapsing under its own weight because it is embedded with mostly very dense radioactive substances.

**White hole:**   When a black hole transforms back to a super monomegastar sometime during the implosion of the universe, then we say a white hole has emerged.

Printed in the United States
By Bookmasters